D1756773

Marine Minerals in Exclusive Economic Zones

Topics in the Earth Sciences

SERIES EDITOR

T.H. van Andel
University of Cambridge

Titles available

Marine Minerals in Exclusive Economic Zones

David S. Cronan

Marine Mineral Resources Programme
Department of Geology

Imperial College of Science,
Technology and Medicine
University of London

CHAPMAN & HALL

London · New York · Tokyo · Melbourne · Madras

Published by Chapman & Hall, 2–6 Boundary Row, London SE1 8HN

Chapman & Hall, 2–6 Boundary Row, London SE1 8HN, UK

Van Nostrand Reinhold Inc., 115 5th Avenue, New York NY10003, USA

Chapman & Hall Japan, Thomson Publishing Japan, Hirakawacho Nemoto Building, 7F, 1-7-11 Hirakawa-cho, Chiyoda-ku, Tokyo 102, Japan

Chapman & Hall Australia, Thomas Nelson Australia, 102 Dodds Street, South Melbourne, Victoria 3205, Australia

Chapman & Hall India, R. Seshadri, 32 Second Main Road, CIT East, Madras 600 035, India

First edition 1992

© 1992 David S. Cronan

Typeset in 10/12pt Bembo by Best-set Typesetter Ltd.
Printed in Great Britain by St Edmundsbury Press Ltd.,
Bury St Edmunds, Suffolk

ISBN 0 412 29270 X

A catalogue record for this book is available from the British Library

Library of Congress Cataloging-in-Publication data available

Contents

Series foreword

Year by year the Earth sciences grow more diverse, with an inevitable increase in the degree to which rampant specialization isolates the practitioners of an ever larger number of subfields. An increasing emphasis on sophisticated mathematics, physics and chemistry as well as the use of advanced technology have set up barriers often impenetrable to the uninitiated. Ironically, the potential value of many specialities for other, often non-contiguous ones has also increased. What is at the present time quiet, unseen work in a remote corner of our discipline, may tomorrow enhance, even revitalize some entirely different area.

The rising flood of research reports has drastically cut the time we have available for free reading. The enormous proliferation of journals expressly aimed at small, select audiences has raised the threshold of access to a large part of the literature so much that many of us are unable to cross it.

This, most would agree, is not only unfortunate but downright dangerous, limiting by sheer bulk of paper or difficulty of comprehension, the flow of information across the Earth sciences because, after all it is just one earth that we all study, and cross fertilization is the key to progress. If one knows where to obtain much needed data or inspiration, no effort is too great. It is when we remain unaware of its existence (perhaps even in the office next door) that stagnation soon sets in.

This series attempts to balance, at least to some degree, the growing deficit in the exchange of knowledge. The concise, modestly demanding books, thorough but easily read and referenced only to a level that permits more advanced pursuit will, we hope, introduce many of us to the varied interests and insights in the Earth of many others.

The series, of which the book forms a part, does not have a strict plan. The emergence and identification of timely subjects and the availability of thoughtful authors, guide more than design the list and order of topics. May they over the years break a path for us to new or little-known territories in the Earth sciences without doubting our intelligence, insulting our erudition or demanding excessive effort.

Tjeerd H. van Andel
Series Editor

Acknowledgements

I would like to express my gratitude to all who have helped me in completing this book. Alan Archer kindly read and commented on Chapter 1, and has always provided sound advice on both the resource potential of marine minerals and the conditions under which they might be mined. Dennis Ardus kindly read and commented on Chapters 2 and 3, provided several useful references and graciously allowed me to quote at length from his paper with D. Harrison on aggregates off East Anglia published in the *Journal of the Society for Underwater Technology*. Many of my case studies are drawn from personal experience, particularly those based on work in the South-West Pacific, and I thank CCOP/SOPAC (now SOPAC) for support in carrying out much of that work. Finally I thank all those authors and publishers who have allowed their diagrams to be reproduced and who are named in the text.

To my wife, Jill Ann, for her continuing
understanding and support.

Preface

In the past decade, the emphasis in marine mineral deposits studies has shifted from the deep sea area to the newly created 200 mile Exclusive Economic Zones, which are under the jurisdiction of adjacent coastal states. This change in emphasis has necessitated a single state-of-the-art account of these mineral deposits, which the present text attempts to supply. However, marine minerals themselves know of no political boundaries and much learnt about them in the deep sea applies equally well in EEZs. It is in their ownership and the conditions under which they can be mined that they principally differ from deep sea minerals more than 200 miles from land. Already we are seeing coastal states, like the United States, enacting legislation to govern the recovery of EEZ mineral deposits, and other countries can be expected to follow suit. Marine mining is already taking place in EEZs, but not yet in the deep sea. It is likely that this situation will not change for at least the remainder of the present century.

David S. Cronan
London
1991

1

Introduction

The oceans comprise just over 70% of the Earth's surface, and the sediments occurring on the ocean floor reflect the history of the ocean basins, and contain mineral deposits of potential economic value.

Little was known about the seafloor, especially in the deep oceans, prior to the *Challenger* Expedition of 1873–6, but that expedition was able to outline the major features of seafloor sedimentation as we now know it. The *Challenger* Expedition was followed by a number of other major expeditions up to the time of the Second World War, and after that ships of many nations worked to provide seafloor sediment and mineral data. Thus we now have an abundance of information with which to assess the economic geology and resource potential of the seafloor.

General recognition of Exclusive Economic Zones as potential locations of marine mineral resources other than oil and gas dates from the early 1980s, and received a considerable fillip with President Reagan's declaration of a US EEZ in 1983. Since that time, many nations have sought to exercise control of the seafloor resources near their coasts. For the purpose of this work, the Exclusive Economic Zone is taken to be that area within 200 miles of a coastal state, in which the coastal state exerts jurisdiction over the resources. World EEZs are shown in Figure 1.1.

In the present work, the minerals dealt with are as follows:

1. aggregates, which are near-shore deposits of non-metallic detrital minerals and calcium carbonate, principally used in the construction industry;
2. placers, which are detrital metallic deposits generally found on beaches and in nearshore areas;
3. precious corals, which are high value corals used for making jewellery;
4. phosphorites, which are authigenic minerals formed *in situ* on the sea floor, in generally deeper waters than placers;

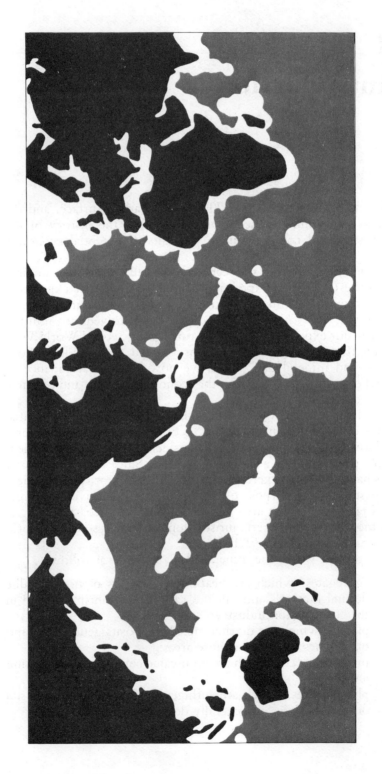

Fig. 1.1 World Exclusive Economic Zones.

5. manganese nodules, which, economically, are deposits principally of manganese, nickel, cobalt and copper, and which occur in the deeper parts of some EEZs, principally around oceanic islands;

6. ferromanganese oxide crusts, the so called cobalt-rich crusts, which occupy seamount areas in and around island chains; and

7. hydrothermal deposits, including polymetallic sulphides, which are precipitates of principally iron, copper and zinc sulphides formed as a result of submarine volcanic activity, but which can also contain enrichments of lead, silver and gold.

Oil and gas are not dealt with at all because there are other texts which adequately cover these deposits.

1.1 PROCESSES OF SUPPLY OF ELEMENTS TO THE SEAFLOOR

In any discussion of the nature and resource potential of marine mineral deposits, some understanding of the processes of supply of elements to the seafloor is needed. For a more comprehensive review of this latter topic see Chester (1990).

1.1.1 Hydrogenous supply

Chemical elements enter seawater in a number of different ways. Sometimes their ultimate source is the dissolved material in rivers draining the continents. In other cases they are derived from hydrothermal solutions entering seawater from submarine volcanoes. Because ocean mixing processes can be slow, elements introduced into seawater from a specific source can sometimes be identified for considerable distances from that source. A good example of this is manganese introduced into seawater at mid-ocean ridges by hydrothermal processes. This can produce a distinctive enrichment of manganese horizontally at the water depth of the crest of the mid-ocean ridge, which can persist for many kilometres away from the ridge, as, for example, on the East Pacific Rise. An example of manganese variability in seawater has been documented by Jones and Murray (1985). These workers collected water samples along a transect near 47°N away from the coast of Washington, in the Pacific North-West region of the USA. They found that manganese concentrations decreased markedly away from the continental margin.

This they considered to be largely due to input of manganese from the Columbia River. However, another source of manganese-rich water near the continental margin was considered to be the shelf sediments themselves. Reducing conditions within the sediments were thought to be leading to the reduction of manganese and its remobilization out of the sediments, and its diffusion into the overlying seawater.

Ultimately, however, ocean mixing and particulate scavenging processes tend to smooth out compositional variations in ocean water due to specific sources. Nevertheless, the oceans are by no means uniform in composition. Some elements exhibit considerable vertical variations in concentration throughout the entire depth of the ocean as a result of their participation in geochemical cycles leading to their extraction from surface and near-surface waters and their reintroduction into seawater at different depths or at the seafloor. Biological removal of elements from surface waters by living organisms or adsorption on to organic material and their reintroduction at a depth at or near the seafloor as a result of the dissolution of sunken organic remains is a good example of such a process. Nickel, for example, is an element whose vertical distribution is influenced by this process (Bruland, 1980).

Manganese also exhibits a vertical variation throughout the ocean. Concentrations are typically high at the surface and low in deep water, although intermediate depth maxima frequently occur. The surface maximum has been related to several possible sources of manganese including river runoff, desorption from particulates, and diffusion from nearshore sediments that are reducing (Bender *et al.,* 1977). There is also sometimes a mid-depth concentration maximum associated with the dissolved oxygen minimum zone (Klinkhammer and Bender, 1980), the latter being the result of sinking organic remains utilizing oxygen in the water column, and, in the Pacific, tends to be most intense at around 1000 m depth. However, it is not uniformly developed throughout the Pacific, being better developed in the North Pacific than in the South (Figure 1.2). Owing to the coincidence of position of the dissolved oxygen minimum and the manganese maximum, an input of dissolved manganese from the degradation of organic material has been proposed as one cause of the excess manganese at this depth. However, it appears that living plankton concentrate only minor amounts of manganese, and that instead manganese is taken up on the surfaces of organic particles by adsorption and is then released back into the water as these particles sink through the oxygen minimum zone. However, an alternative explanation of much of the manganese enrichment at the depth of the

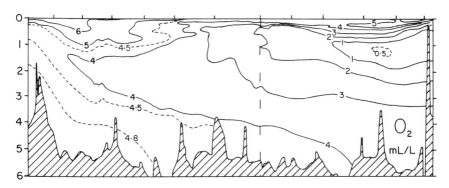

Fig. 1.2 Distribution of dissolved oxygen in the central Pacific Ocean (from Pickard, 1975).

dissolved oxygen minimum is a horizontal flux of manganese out of continental margin sediments. The importance of this in nearshore areas has been mentioned above, but Martin and Knauer (1984) consider that this process also affects manganese distribution in the deep ocean, and indicates the importance of horizontal advective-diffusion processes in influencing the chemistry of the oceans.

Removal of elements from seawater and from the interstitial waters of sediments can take place by a number of reactions. These include direct precipitation (as in the case of evaporites); catalytic precipitation (as in the case of manganese oxides); and adsorption (as in the case of a host of minor metals). The phases formed by direct removal of elements from seawater into sediments are termed hydrogenous phases, and because they generally tend to accumulate slowly they reach their greatest concentrations in sediments where the supply of other phases is limited. Also generally included as hydrogenous phases are those formed as a result of the alteration of pre-existing minerals by seawater. These are termed halmyrolysates (after halmyrolysis). Halmyrolysis is a process embracing chemical and physico-chemical processes that occur as a result of reactions between sediment components and seawater (Elderfield, 1976). Examples include the alteration of minerals in the suspended loads of rivers when they enter the sea, and the alteration of submarine volcanic assemblages. The latter is particularly important in areas where volcanic rocks outcrop and can markedly influence the bulk composition of the sediments in those areas. Of the minerals dealt with in this work, manganese nodules and cobalt-rich crusts are primarily hydrogenous in origin.

1.1.2 Lithogenous supply

The lithogenous (detrital) fraction of marine sediments consists of material delivered to the oceans as solids which undergo little alteration during their transport and final deposition (Windom, 1976). In a classification erected by Goldberg (1954), lithogenous components of marine sediments were subdivided into those derived from land erosion and submarine volcanic activity. Land-derived lithogenous material is termed terrigenous material and submarine volcanic lithogenous material is termed volcaniclastic material.

Terrigenous materials are the solid products of continental weathering and their composition reflects that of the continental rocks from which they were derived. They enter the oceans in a particulate form, and remain in that form throughout their depositional processes. Clay minerals, quartz and feldspars are the most important lithogenous components in marine sediments. Clay minerals are aluminosilicates, and of all the elements contributed to marine sediments by terrigenous supply, aluminium and silicon are the most abundant, with magnesium, iron, potassium, and titanium of lesser importance. Of the minerals dealt with in this work, placer minerals and aggregates are primarily lithogenous in origin.

Volcaniclastic materials are the solid products of volcanic activity, both terrestrial and submarine. They occur in marine sediments largely in the form of fine ash and volcanic glass, although sometimes larger and more exotic volcanic products such as volcanic bombs may occur. Because basic rocks are far more abundant than acid and intermediate rocks in and around ocean basins, volcaniclastic materials tend to approximate to a basaltic composition. Thus they are rich in iron and magnesium but are poorer in aluminium and silica than many terrigenous materials.

1.1.3 Hydrothermal supply

In recent years, it has been realized that hydrothermal supply of elements to the oceans is an important contributor to the composition of some mineral deposits.

As a result of submarine volcanic activity, new seafloor is created at mid-ocean ridge crests. Seawater enters cracks and fissures in the hot, newly formed rocks, becomes heated and engages in chemical reactions with them (Figure 1.3). This leads to the leaching of iron, manganese, copper, zinc, and other elements out of the rocks, and the reduction of seawater sulphate to sulphide. Research has indicated a possible seawater interaction with volcanic rocks down to

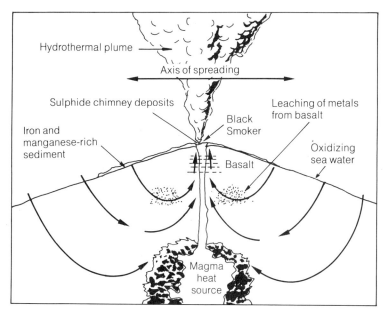

Fig. 1.3 Sub-sea floor leaching processes and black smoker formation (modified from Cronan, 1985).

a depth of several kilometres below the seafloor. The nature of the reactions that take place has been verified in the laboratory by reacting seawater with mid-ocean ridge type basalt at elevated temperatures (Seyfried and Bischoff, 1977). The ratio of seawater to unit volume of rock leached has a profound effect on the composition of the resulting hydrothermal solution.

After reacting with newly formed ocean-floor rocks, the hydrothermal solutions ascend to the seafloor, and precipitate at least a fraction of the metals they have obtained as sulphide minerals and metalliferous sediments. The generalized precipitation process on cooling of the hydrothermal solution is as follows: first, the deposition of sulphide minerals, principally of iron, copper and zinc; second, the precipitation of silicates sometimes by the reaction of silica with iron oxides; third, the precipitation of iron oxides; fourth, the precipitation of manganese oxides. The spatial relationships on the seafloor between these different deposits depend largely on the physicochemical conditions at and near the point of hydrothermal discharge. Discharge of hydrothermal solutions into normal seawater is likely to lead to rapid precipitation of most of the hydrothermal constituents with the formation of 'black smokers' (see p. 135) rich in copper, zinc, and iron sulphide particulates and the growth of sulphide-rich 'chimneys'.

These deposits may be surrounded by a halo of iron-rich sediments, and a plume of iron and manganese rich particulates can develop in the overlying seawater which can extend and precipitate oxides for many kilometres away from the hydrothermal source (Figure 1.3). If picked up by ocean currents, the hydrothermal iron and manganese oxides can precipitate as much as hundreds of kilometres away from their point of discharge, as happens on the flanks of the East Pacific Rise for example (Edmond *et al.,* 1982). If, by contrast, the hydrothermal solutions discharge into an environment in which the supply of oxygen is limited or absent, as for example in some of the deeps of the Red Sea, initial separation of the hydrothermal precipitates will be more pronounced as the physicochemical properties of the depositional environment change more slowly than in the open ocean, and sulphides, silicates, and iron oxides are likely to be more widely separated (Cronan, 1980). Should the hydrothermal solutions cool sufficiently below the seafloor, major sub-seafloor precipitation of hydrothermal minerals will take place there.

1.1.4 Cosmogenous supply

The possibility that cosmic supply of elements to the seafloor might be important, at least for a few elements such as nickel and platinum, has been discussed from time to time (Petterson and Rotschi, 1952; Agiorgitis and Gundlach, 1978). However, while cosmic spherules are present in marine sediments, their abundance is low, and they do not contribute substantially to the bulk composition of any marine mineral deposits. Nevertheless, it is possible that platinum of cosmogenic origin may be contributing to the platinum enrichments in some cobalt-rich crusts (Wiltshire, personal communication, 1990).

1.1.5 Biogenous supply

Biological supply of elements to the seafloor can be brought about in several ways, some direct and others indirect. The most direct is the sinking of the remains of organisms, and the liberation of the elements they contain as they decay. Calcareous ($CaCO_3$) and siliceous (SiO_2) organisms are the two most abundant varieties of plankton in surface waters, and their remains constitute the bulk of biogenous sediments on the ocean floor. Other ways in which elements can be biologically supplied to marine sediments are as part of the life processes of organisms. Elements are extracted from seawater as part of the feeding process of planktonic organisms, pass up the food chain and are excreted in faecal pellets that fall to the seafloor. These and other organic materials can scavenge additional amounts of elements from the seawater through which they fall, thereby indirectly enhancing the biological flux of metals to the seafloor.

Of course, biological supply of elements to the seafloor cannot be considered a primary source of elements to sediments or mineral deposits as the elements must be present in seawater already. Rather, biological processes provide a mechanism by which elements in seawater are concentrated prior to their removal from the oceans. Phosphorites are seafloor mineral deposits showing a major biogenic influence, manganese nodules show some biogenic influences, and of course precious corals are wholly biogenic.

1.2 LAW OF THE SEA

Man's desire to exert a legal control over resources on the ocean floor can be said to date from the end of the Second World War, when President Truman proclaimed exclusive jurisdiction and control for the United States over resources on its continental shelf. Prior to this, the continental shelf beyond the territorial limit had been regarded as part of the 'high seas', which belonged to no one. Other states followed the lead of the United States, and in 1958 the Geneva Convention on the Continental Shelf, gave coastal states 'Rights' over their continental shelf resources.

During the following decade. the mineral potential of the deep sea floor beyond the continental margin (Figure 1.4) came to be recognized, mainly as a result of the pioneering work on manganese nodules carried out by John Mero (Mero, 1965). Under the terms of the 1958 Geneva Convention, the outer limit of the Continental Shelf was taken to be at a depth of 200 m, or beyond that limit to a depth that could be exploited. This meant, in theory, that there was no limit on a coastal state exploiting the deep sea floor adjacent to its coast so long as it had the technology to do so.

The increasing awareness during the 1960s that the technologically advanced states could exploit that deep sea floor for their own benefit was at least partly instrumental in leading to the advocacy by Arvid Pardo of Malta that the sea floor beyond the limits of national jurisdiction, together with the resources on and under it, should be considered the common heritage of mankind. In other words, no

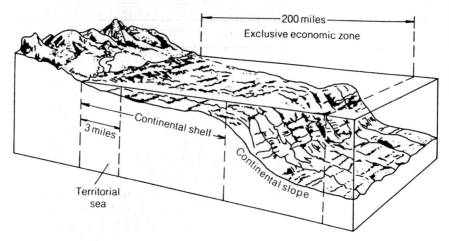

Fig. 1.4 Physiography of the continental margin (from McGregor and Lockwood, 1985).

state or multinational corporation should be allowed to exploit it for its sole benefit. This advocacy led to the Third United Nations Conference on the Law of the Sea (UNCLOS III) which concluded in 1982, and which incorporated the common heritage principle into its deep seabed mining provisions.

The UN regime proposed by UNCLOS III for the seabed more than 200 miles from land (known as the Area) contains a number of provisions. These include:

1. the 'common heritage' principle mentioned above;
2. a parallel system of deep sea mining under an International Sea-Bed Authority which could mine the sea floor itself through a subsidiary called the Enterprise, and license other organizations (states or multinationals) to mine it also;
3. the transfer of seabed mining technology from technologically advanced nations to less developed seabed mining nations; and
4. production controls on seabed minerals to protect land based producers.

In addition, Conference Resolutions 1 and 2 respectively established a Preparatory Commission to act until the convention came into force following 60 ratifications, and the PIP scheme which gave protection to certain 'pioneer investors' who had already expended considerable sums on deep sea mineral exploration. Under the latter scheme, a mining consortium could obtain a preferential status by being registered as a pioneer investor after the fulfilment of certain conditions, and would have a 'pioneer area' allocated to it. At the time of writing, there are four pioneer investors registered with the Preparatory Commission, mining ventures of Japan, France, the USSR and India.

Several of the technologically developed countries objected to the deep seabed mining provisions of the Law of the Sea Convention, and some voted against it. These included countries which had been involved in multinational deep sea mining consortia such as the USA, the UK and the FRG. These and other countries had already established deep sea mining regimes based on their national legislation which respected each other's rights and which was to apply until the UN Law of the Sea Convention came into force for them. Several mine site claims were made in the 1970s and 1980s, some of which in the Pacific overlapped. These overlaps have now been resolved. The only mine site claimed outside the Pacific Ocean was one by India in the Indian Ocean.

As far as marine mining is concerned, another important consequence of the Third Law of the Sea Conference was the general acceptance of the 200 mile Exclusive Economic Zone which had previously been unilaterally claimed by some states but which had not been recognized by many others. The regime for the continental shelf established under the 1958 Geneva Convention was more or less incorporated into the 1982 Law of the Sea Convention. The EEZ concept is generally recognized even by states that have not signed the Law of the Sea Convention. Declining prices for the minerals in the deep seabed Area coupled with perceived difficulties in mining there have served to shift the emphasis in marine mineral exploration in recent years from the Area to the EEZs.

1.3 MINERALS FOUND IN EEZs

Exclusive Economic Zone minerals have a potential to occur in any of the physiographic settings occupied by seafloor minerals in general (Figure 1.5) (see Cronan, 1980 for a review). Continental margins are the principal sites of aggregates and placers, commercially interesting examples which tend to be preferentially located in their inner portions and on beaches. Phosphorites are also typically continental margin deposits, but generally occur in the deeper areas relative to placers and aggregates. EEZ manganese nodules of potential economic interest occur exclusively seawards of continental margins, generally on the abyssal seafloor adjacent to oceanic islands.

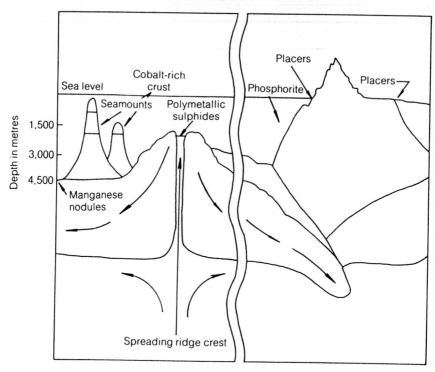

Fig. 1.5 Physiographic setting of sea floor mineral deposits (modified from McGregor and Lockwood, 1985).

Cobalt-rich crusts occur on seamounts and to a much lesser extent on the upper submerged slopes of oceanic islands, and are characteristically most abundant far from major land masses. Hydrothermal deposits have the potential to occur in any areas of submarine volcanic activity and have been recorded to a greater or lesser extent on mid-ocean ridges, intra-plate hotspots, and at convergent plate margins including both island arcs and back-arc basins. However, current data show them to be most abundant on mid-ocean ridges, but this may simply be a reflection of the more limited sampling in other submarine volcanic areas.

1.4 RESOURCE CONSIDERATIONS

What actually constitutes marine mineral resources has, until recently, defied strict definition. According to Archer (1987), the subject has been fraught with confusion, with several different definitions of reserves and resources having been used in the past. In an attempt to clarify the situation, Archer (1987) proposed that marine mineral resources should be defined as mineral occurrences below the sea which are either economically workable now or are likely to become economically workable within the next 20 or 30 years. On this basis, aggregates, placers and precious coral can be considered to be present-day marine mineral resources as currently all of them are being recovered. It will be shown later that some hydrothermal deposits, manganese nodules and phosphorites have all been projected as being exploitable during the next 30 years and thus these too can be considered to be marine mineral resources in the sense defined by Archer (1987). The only seafloor mineral deposits that would not appear on the basis of present evidence to fall within Archer's definition of marine mineral resources are cobalt-rich crusts.

Notwithstanding the above, mankind's perception of the resource potential of marine mineral deposits has undergone considerable changes over the past 30 years. This has been mainly exemplified by assessments of the worth of manganese nodules, but to a certain extent perceptions of the value of other seafloor minerals have also varied.

Although known since the latter part of the nineteenth century, it was not until the post-war period that serious consideration was given to manganese nodules as a possible mineral resource (Mero, 1965). They were regarded, and still are, primarily as a source of nickel, copper and cobalt, rather than manganese and iron, the two most abundant metals in them, although some extraction scenarios also envisage the recovery of manganese. However, in addition to their potential as a source of metals, they may also find applications in pollution control. It has been found that nodules can absorb up to 200% of their weight of sulphur dioxide, a common gaseous effluent in power plants, and thus they may find applications in the control of gaseous emissions as a stack gas decontaminant (Siapno, 1989). Serious mining company interest in manganese nodules commenced in the middle 1960s and seemed to peak about 1981–2. Static or declining metal prices, until recently, were largely the cause of a diminution of interest in nodules by mining companies, but perceived difficulties in nodule mining under the Law of the Sea Convention have also served to inhibit nodule development. Indeed, what is widely regarded as a restrictive regime for deep sea mineral exploitation proposed in the Law of the Sea Convention can be held partly responsible for increased interest in manganese nodules in EEZs in the post-1982 period. Recent increases in metal prices have started to revive mining company interest in manganese nodules, and indeed the proposed French nodule project is projected as being profitable based on 1980s metal prices (Herrouin *et al.*, 1989).

Economic interest in other sea floor minerals has not followed such a pendulum-like motion as that in manganese nodules. Placers have been mined intermittently for many years and still are being mined in various parts of the world today. Aggregates are an important source of materials for construction near the coast, and indeed in recent years have been transported to more inland sites if local supplies of aggregate are limited there. Until recently, little serious attention has been given to the possibility of marine phosphorites as a resource, largely as a result of the large quantities of phosphorite on land. However, it is becoming recognized that some of the areas where phosphorites occur on land have other potential uses than simply for phosphorite mining, i.e. building, recreation, etc., and therefore access to them may be restricted in the future. This would work to the advantage of seafloor phosphorite extraction by increasing the price of phosphate. Interest in cobalt-rich crusts entirely post-dates UNCLOS III and is held by some to be a good example of a reaction to it. Interest in hydrothermal deposits is also relatively recent, and first emerged in relation to the Red Sea hydrothermal deposits in the 1970s. Since that time, many more hydrothermal deposits have been discovered in submarine volcanic areas around the world. Nevertheless, the greatest present economic interest in mining hydrothermal deposits remains in the Red Sea.

It is difficult to make firm predictions concerning the pace of development of EEZ mineral deposits extraction, or indeed the extraction of sea floor mineral deposits in general. As the demand for raw materials grows, seabed deposits can be expected to be increasingly developed. However, as Broadus (1987) points out, the potential seabed mineral resources will have to be viewed in relation to both conventional and more speculative rival mineral resources on land, or wait to be developed when little or nothing of those resources remains. Broadus (1987) considers the best approach to the identification of various seabed minerals as emergent resources is a long run supply function for each mineral. These are based on the amounts of each mineral that could be obtained economically at different levels of incremental or unit cost. For most seabed minerals, one would expect continuing onshore production from successively costlier deposits until they become so expensive that the least cost seabed equivalents could be mined in competition with them. Once that threshold has been reached, the division of output between onshore and offshore mineral resources would depend on their respective available quantities for each increment of elevated cost.

Several positive attributes of EEZ minerals relative to those in the international area have been cited, as follows:

1. their security of ownership by the coastal state;
2. their general nearness to land; and
3. their generally shallow depth.

In addition, reducing imports of strategic minerals has been cited as a reason for developing such minerals in those Exclusive Economic Zones where they occur. In President Reagan's 1983 EEZ Declaration, he referred to 'recently discovered deposits' that might be an important future source of strategic minerals.

In the remainder of this work, each of the minerals introduced in this first chapter and the concepts touched upon, will be explored in greater detail on an individual basis.

2

Aggregates

Aggregates are non-metallic deposits consisting of sands, gravels, shells or coral debris, which have their main use in the construction industry. They occur both on beaches and in the offshore area, and all are EEZ deposits. Marine aggregates have been concentrated into their present occurrences by seafloor hydrodynamic processes although their deposition may originally have been by some other mechanism; by rivers or glaciers for example. Quantitatively they are the most important offshore mineral deposits being extracted at the present time, if one does not consider oil and gas to be mineral deposits. A classification of marine aggregates on the basis of grain size is given in Table 2.1.

2.1 SAND AND GRAVEL

Gravels generally consist of stable rock fragments such as quartzite, chert or flint which are transportable for considerable distances without breaking down. Stable minerals also generally comprise the bulk of aggregate sands, with quartz usually being dominant. However, non-quartzose aggregate sand deposits do occur (Cronan, 1980) and calcareous sand is particularly common on tropical beaches.

On modern beaches, the erosion of local sources of aggregate material such as cliffs, provide much of the aggregate on the beach. However many beaches also accrue aggregate material from offshore as a result of onshore wave and current transport. Conversely, aggregates in offshore areas sometimes consist of transported material from the adjacent coast, but many aggregate deposits on continental shelves are the reworked products of older deposits laid down during glacial periods when sea level was lower.

Aggregate deposits on middle and high latitude continental shelves often result from the reworking and redistribution of glacial and fluvioglacial deposits laid down during past ice ages. When the ice

Table 2.1. Classification of aggregates by grain size

Size limit	Grain size description	Qualification	Primary classification
	Cobble		
64 mm			
		Coarse	Gravel
16 mm	Pebble		
5 mm (2 mm in some classifications)		Fine	
		Coarse	
1 mm			
	Sand	Medium	Sand
0.25 mm			
		Fine	
0.063 mm			
	Fine (silt + clay)		Mud

sheets melted, they deposited a mixed assemblage of minerals and rock fragments to form boulder clay, or fluvioglacial deposits of sand and gravel discharged from glacial melt water streams. During the subsequent rise in sea level, this material was reworked, firstly in beach zones, and later by wave and tidal current action on continental shelves. This reworking has significantly influenced the distribution of many aggregate deposits on middle and high latitude continental shelves such as those off north-western Europe and the north-eastern United States and Canada.

Aggregate transport and redistribution on continental shelves is largely brought about by wave and tidal current action. The amount of material transported is a non-linear function of the strength of the current. The strongest tidal currents are thus a principal transport mechanism, and in these cases net sediment movement can be determined by only slight differences between ebb and flood tide velocities. In addition where wind-induced currents enhance the strength of the tidal current, the amount of sediment that can be moved can be considerably increased. Peak tides augmented by storm generated currents can move the greatest amounts of sediment on the open continental shelf.

The actual distribution of aggregate deposits offshore reflects the original distribution of relict material and its redistribution under the energy level of the depositional environment. For example, gravels can occur as lag deposits where tidal scour is sufficiently strong to remove finer grain sizes from a mixed deposit. Many gravel deposits on glaciated continental shelves may have originated in this manner.

Aggregate sands can also be relict in origin, but most are probably formed by redistribution of previously deposited material. The removal of sand from a mixed gravel and sand deposit and its transport by slackening tidal currents will result in a grain-size separation in a down-current direction, finer and finer grain sizes being deposited as the tidal current velocity decreases (Stride, 1963).

2.2 CALCAREOUS DEPOSITS

Calcareous aggregates can comprise either shells, whole or broken, or finely comminuted reef debris. The latter of course only occur in low latitude areas. Shell deposits tend to be concentrated by hydrodynamic processes on the sea floor and to be effectively separated from non-calcareous aggregate materials such as quartz, sand and gravel (Cronan, 1980). Shells have been mined off Iceland for the production of Portland cement and there are extensive deposits off the northern and western coasts of Scotland (Ardus, 1990 personal communication). However, they are the principle aggregate material in some tropical areas where no quartzose material occurs on the sea floor (O'Neill and Woolsey, 1988) (see south-west Pacific aggregates case study p. 24).

2.3 CASE HISTORIES

Two case studies have been selected to illustrate the diversity of aggregate deposits: one in a well studied, and mined, area off north-western Europe, and the other in a relatively underdeveloped area for offshore aggregates, the south-west Pacific.

2.3.1 Aggregates in the southern North Sea

In the south-east of England, on average 25% of the demand for raw materials for the construction industry is met from marine dredged sand and gravel (Harrison and Ardus, 1990). Future supplies of marine dredged aggregates will depend on the identification of resources and the licensing of reserves of marine sand and gravel. Whereas a significant database exists on the extent of land resources of sand and gravel in the UK, information on the distribution, quality and quantity of marine aggregate resources is sparse.

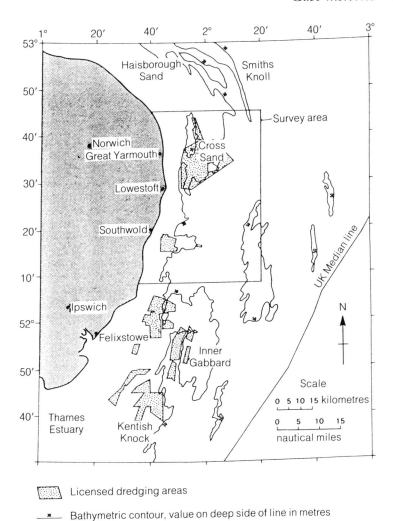

Fig. 2.1 British Geological Survey study areas for aggregates in the southern North Sea (from Harrison and Ardus, 1990).

In 1986–7, the British Geological Survey carried out a survey off East Anglia (Figure 2.1), which was intended to provide a reconnaissance investigation of marine aggregate resources, and aimed to develop a knowledge of the regional geology in order to predict the occurrence and distribution of marine sands and gravels (Harrison and Ardus, 1990). This work is summarized and extensively quoted below and is in an area important for aggregate supply to both the UK and the Netherlands.

Seismic profiling allowed the stratigraphical relationships of individual sedimentary units to be identified and interpreted to give a three-dimensional picture of the regional geometry of each unit. Shallow cores and surface grab samples provided the necessary ground-truth data to calibrate the geophysical interpretation and obtain material for laboratory testing.

The geophysical survey was based on a regular 5 km × 5 km grid of traverses, with more closely spaced lines at 1.5 km intervals. Seabed samples were obtained using a clamshell grab of 0.65 m^3 capacity. Most of the uppermost sedimentary units were sampled using a BGS vibrocorer fitted with a 6 m barrel of 10 cm or 15 cm diameter. Penetration was variable, up to a maximum of 4 m, but in most cases it was of the order of 1.5–2.0 m. Additional seabed data were obtained using an underwater camera system.

The size range of each bulk sample and of selected core samples was determined by sieve analysis in order to obtain the relative proportions of gravel, sand and fines in the sediment and to classify the marine aggregate resources (Table 2.1).

The carbonate contents of both gravel and sand fractions were determined; the former by hand picking the shell from the lithic gravel, the latter by acid digestion.

The early Pleistocene stratigraphic succession in the southern North Sea reflects the expansion of deltaic sedimentation from the Rhine, Meuse and North German River systems across what is now the Netherlands and on to the continental shelf area. Seismic evidence suggests that deltaic sedimentation had extended across the entire southern North Sea by Middle Pleistocene times. The Lower and early Middle Pleistocene sediments are therefore variable in seismic and sedimentary facies and represent a range of marine and non-marine sedimentary environments.

A late Middle Pleistocene unconformity separates the early Pleistocene delta-related sediments from a succession of periglacial, marine and brackish-marine deposits of late Pleistocene to Holocene age. Three regional glaciations affected north-west Europe, although in the late Pleistocene the ice margins lay north of this part of the southern North Sea. There are no glacial deposits (till) in the Pleistocene sequence of this area although till is at or close to the seabed over large areas to the north of East Anglia. The late Middle and Upper Pleistocene succession consists of shallow marine and brackish-marine sandy clays which were deposited during the marine regression at the onset of the most recent, Weichselian, glaciation. Sea level continued to fall until the late Weichselian, when periglacial conditions prevailed and the area emerged as dry land.

At the beginning of the Holocene epoch, global sea level was several tens of metres below its present level but as the ice sheets melted, sea level began to rise and the earliest marine incursion into the southern North Sea occurred before 9000 years BP. With continuing sea level rise the environment became fully marine after c. 7000 years BP. The early Holocene shoreline lay to the east of the present East Anglian coast and the land surface was subjected to the processes of erosion and deposition by river systems which drained the area. During the flooding of the area these fluvial deposits, and older Pleistocene sediments, were reworked to form beach deposits and related near-shore deposits. These were later modified into, or covered by, sheet sands deposited by tidal currents. By approximately 5000 years BP, sea level had stabilized near to the present level and the tidal current system was established. Repeated reworking of the seabed sediments by tide and wave action has facilitated a gradual removal of the fine sand fraction. The superficial (Holocene) sediments are relatively thin, reflecting the small input of sediment into the southern North Sea during the last 10 000 years. Present-day supply of sediment into the area is limited. Erosion of the Pleistocene sediments of the Norfolk cliffs provides an important source of sand-sized sediment to the area, but the contribution from rivers is small.

The present-day floor of the survey area is extensively covered by sands and gravelly sands. The thickness of the superficial sediments, interpreted from seismic and sample data, is usually less than 1 m. Sediment thickness is more variable in some areas and particularly thick accumulations of predominantly sandy sediments are locally developed in nearshore areas off Great Yarmouth, Lowestoft and Southwold.

The distribution of the superficial sediments results from conditions prevailing since the Holocene transgression. At the end of the Pleistocene epoch, the area formed part of a low-lying land surface drained by a system of easterly or north-easterly flowing rivers. As the sea level rose at the beginning of the Holocene, the fluvial sediments were reworked by the advancing sea and re-deposited to form the present distribution of superficial sediments. The gravels are therefore relict deposits reworked only to a minor degree. Recent deposition reflects the tidal current pattern in the North Sea which was established approximately 5000 years ago. Since then, the surface sediments have been sorted to varying degrees and sand has accumulated locally forming sand wave fields. Long-shore drift has locally (e.g. south-east of Lowestoft) resulted in thick accumulations of sand overlying older Holocene gravelly sands.

Aggregate sand resources are extensive throughout the area, occurring in both superficial sediments and in the underlying Pleistocene sequence. Gravelly sediments are mainly restricted to the seabed sediments of Holocene age and are much less common in the Pleistocene strata.

Over most of the area, flint comprises over 90% of the gravel. Phosphorite of Tertiary age and quartzite are also locally important. Most of the flint is of pebble size (<64 mm) and only a small proportion of the pebbles are larger than 32 mm. The seabed is particularly rich in gravel in two large areas off Great Yarmouth and Southwold. In these areas, the surface layers typically contain up to 60% gravel, although the sub-surface layers (below 30 cm) contain a larger proportion of sand and typically 15–30% gravel. The concentration of gravel in the surface layers is due to the winnowing of sand and finer sediment by tidal currents. The proportion of shell in both gravel and sand fractions of the sediment is rarely greater than 10%.

Sand-sized sediment in both the superficial and Pleistocene deposits is mainly of medium grain size and most occurs in the 0.25–0.50 mm size range. Muddy sediment at the seabed is found only in nearshore areas between Southwold and Aldeburgh. The underlying Pleistocene strata, however, are dominated by clays and muddy sands which cover large parts of the study area, often within 0.5 m of the seabed. In some areas the veneer of potential marine aggregate resources may be locally contaminated by these underlying deposits of clays and muddy sands.

In conclusion, the distribution and potential quality of marine sand and gravel resources off Great Yarmouth and Southwold has been clearly identified by the BGS research programme. The major gravel resources are found within the superficial (Holocene) sediments and are relict deposits resulting from the reworking of former fluvial sediments during the Holocene transgression. Flint is the dominant lithology. Sand-sized sediment is common in the underlying Pleistocene sequence and gravel is subordinate there, except for a few places where there are potential gravel resources (Harrison and Ardus, 1990).

The survey has shown that most of the area is underlain by Eocene and Pleistocene clays and muddy sands concealed beneath a thin cover of superficial sediments. This information will be useful to the dredging industry to help avoid contamination of the aggregates by muds and clays.

2.3.2 Offshore aggregates in the south-west Pacific

Whereas the greatest demand for marine aggregate is in the heavily urbanized coastal areas of the northern hemisphere, there is an increasing need for the deposits in some tropical areas. The main reasons for this are, first, urbanization of coastal areas, and second, tourism, resulting in large scale building projects such as hotels and airports. Much of the aggregate needed for construction purposes in the coastal areas of large tropical land masses such as Australia or India, can be derived from beaches and inland sites. However, the aggregate requirements of small tropical islands often cannot be met from these sources. The reasons for this are, first, the small size of the hinterland, and second, the recreational value of the beaches.

Good examples of some of the problems encountered in tropical aggregate development have been those experienced by some of the countries in the south-western Pacific (Figures 3.15 and 4.1) (Gauss, personal communication, 1984).

In the larger, high volcanic island countries, such as Fiji, Vanuatu, the Solomons and Papua New Guinea, there has been little or no problem with aggregate supply. Stone is quarried and crushed for coarse aggregate, with some fine material as a by-product. Building sand is often taken from river deposits or from beaches that are adequately replenished with material brought down by the rivers.

It is in the low lying coral atolls and coral islands, and in some of the small volcanic islands, that the problem of aggregate supply is sometimes causing concern. Removal of material from the reef is sometimes a short-term solution to the aggregate problem in such areas, but can lead to enhanced coastal erosion and possibly to the ultimate destruction of beaches. Removal of aggregate from the beaches themselves is ultimately self-defeating as it is just these beaches that are often a main tourist attraction. Admittedly beach material may be replenished over a long period of time by abrasion of the coral reef supplying foram tests, sponge spicules, Halmeda remains, etc. and by comminution of broken shell material, but this is a long-term process and cannot be expected to replenish beaches at the rate they would need to be mined in order to produce significant aggregate supply. In certain cases in the south-western Pacific, entire beaches have been removed, initially by mining, and subsequently by enhanced coastal erosion partly consequent on the mining. Problems such as these have led to an increased effort in the search for offshore aggregates in the south-western Pacific over the past few years.

According to O'Neill and Woolsey (1988), the backreef coral aggregate which occurs in the nearshore region of virtually every island nation in the south-western Pacific is a material of particular interest for industrial development. Offshore coral aggregate has potential for use either as a construction aggregate for buildings and roads or for the manufacture of cement. Typically, it is composed of unconsolidated skeletal debris that accumulates in draping strata on the landward side of a reef. Wave action and the work of biota are responsible for breaking off portions of the reef, leaving the coral debris to accumulate. Particle sizes in the backreef facies are typically sand to gravel, with an additional silty fraction made up of organic-calcareous matter.

Of the south-western Pacific nations, Western Samoa has taken advantage of offshore aggregate for some time. Close to Apia there is an extensive shallow water lagoon enclosed behind a small peninsula and the fringing reef which is here about half a mile in width (Gauss, personal communication, 1984). This lagoon appears to have been accumulating reef debris for a considerable period of time (possibly it is sinking slowly) and now is floored by an unsorted deposit of sand, broken coral and small coral boulders which is at least several metres thick. The technique in Apia for mining this deposit has been to build narrow causeways out into the lagoon, by dredging up the lagoon bed using a dragline dredge, and dumping it as required. The same material has then been further dredged by the dragline and dumped in piles on the causeways to be transported back to the shore where it has been screened by a standard mechanical screener. This has produced a sand with a very good size distribution for concrete making, an aggregate of broken coral pieces about 3 cm across for general building purposes and roads and, finally an aggregate of the larger coral pieces for 'heavy' construction purposes.

According to Richmond (1990), additional Western Samoan aggregate sources consisting of mixed carbonate and terrigenous materials are concentrated in drowned river valleys behind active reefs. The deposits are thought to have been trapped in river valleys during the Holocene sea level rise by barrier reef growth that prevented their seawards transport.

The island of Tongatapu in the Kingdom of Tonga is surrounded by a fringing reef which provides the source for sand forming many of the island's beaches. The area of the fringing reef has been estimated to be about 3 462 000 m^2 and is capable of producing 3462 tonne/year (Kitekei'ahu and Harper, 1988). It is estimated that the fragmentation of the reef front and transport of fragments to the beaches only amounts to about half this amount (i.e. 1731 tonne/

year). The 1988 rate of beach mining was estimated at about 15 000–20 000 tonne/year. As this rate clearly exceeds the production rate of the reef, it would have depleted the beach sand resources and caused coastal erosion problems. As a result of this, alternative nearshore aggregate sources were sought. Abundant backreef carbonate deposits occur throughout Tonga's broad reef platform (Richmond, 1990). Surveys near Tongatapu and Vava'u, an island to the north, have revealed several areas with potentially large reserves of carbonate sand. Early estimates for an area near Fafa Island off Tongatapu indicate at least 3 million cubic metres of sand. According to Richmond (1990), the presence of widespread ripple fields suggests that the distribution of the deposits is current controlled.

In Funafuti Atoll (Tuvalu) aggregate is needed to fill pits dug during the Second World War in order to provide airfield construction materials. Richmond (1990) has reported extensive deposits of Halmeda sand in the Funafuti lagoon. Additional information on aggregate deposits in south-western Pacific EEZs is given in Glasby (1986).

2.4 RESOURCE POTENTIAL OF MARINE AGGREGATES

The resource potential of marine aggregates can be said largely to depend upon the economics of mining the material offshore, relative to that of mining it on land. This partly depends on the grade and quality of the deposits, and on their distance from potential markets as land transport costs for bulky aggregates are high with prices doubling with transport over 24 km (15 miles). Other factors such as the possibly deleterious environmental impact of aggregate mining onshore, and alternate land use scenarios such as building and recreation, and increasing land prices, all work in favour of offshore aggregate mining.

Offshore aggregate is of major importance in some heavily urbanized areas close to the sea. Examples include south-eastern England and Japan. These are the areas where most work on offshore sand and gravel extraction has been done. In 1987, 16 million tonnes of sand and gravel were mined from the UK continental shelf out of a total UK production of more than 100 million tonnes, and by 1988 marine aggregate had grown to 16% of the total amount of sand and gravel production in England and Wales (Ardus and Harrison, 1989). However, Japan has overtaken the United Kingdom in offshore aggregate production and now produces more than 50% of the

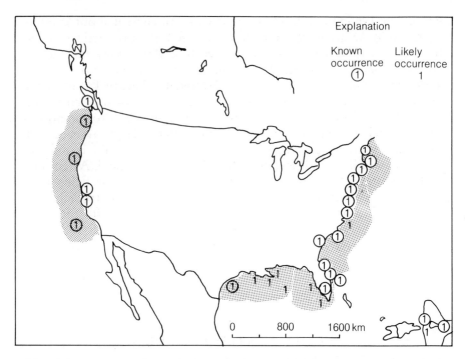

Fig. 2.2 Aggregate deposits along the coasts of the United States (modified from US Congress, Office of Technology Assessment, 1987).

world's offshore aggregate supply, over 70 million tonnes per year (Tsurusaki *et al.*, 1988).

In the United States, rising aggregate consumption and depletion of available resources on land have reduced available sand and gravel in and near many urban areas. Another problem faced by sand and gravel producers is that deposits close to urban areas are often being removed from potential production by suburban growth and other alternate land use. Increasing environmental awareness is also acting to the detriment of onshore sand and gravel production. Increasingly, environmental controls are being placed on sand and gravel extraction; production is often restricted, and sometimes environmental considerations have precluded sand and gravel mining altogether. Factors such as these are contributing to an increasing interest in offshore aggregate production in the United States.

Off the USA, there are large aggregate deposits (Figure 2.2). It has been estimated that off the coasts of the United States there are over

15 billion cubic yards or 21 billion tonnes of usable sand and gravel within 5 km (3 miles) of the coast (US Congress, 1987). In a study on the economic aspects of offshore sand and gravel mining in the United States, Dehais and Wallace (1988) conclude that offshore mining of fine mineral aggregate could be sufficiently profitable to induce investment, utilizing some of the resources just mentioned. However, until the legal, tax, political and regulatory concerns associated with offshore mining are resolved, investment is not likely to occur.

In Japan, rapid depletion of sand and gravel in riverbed deposits has led to mining sand from deposits in coastal waters. In recent years, such seabed mining has contributed 20–25% of the total production of natural aggregate and about 10% of all aggregate used in Japan (Tsurusaki *et al.*, 1988). The major seabed sand mining occurs off western Japan (Figure 2.3). Sixty per cent of the production occurs in the Inland Sea, and an additional 35% occurs off the north coast of Kyushu and the south coast of Shikoku. The total number of mining vessels was about 540 in 1988. In recent years, environmental impact and the depletion of resources in shallow water have caused operators to move the mining to a greater distance from the shoreline and into deeper water (Tsurusaki *et al.*, 1988). In 1989, total Japanese aggregate production was worth approximately 100 000 000 000 yen (Arita, personal communication, 1989).

In the United Kingdom, the development of the offshore sand and gravel industry during this century provides an interesting example of the increasing importance of offshore aggregates. In the early days, the aggregate was dredged with grab cranes mounted on board extraction vessels, but these were only able to operate at shallow depth and in good weather. However, after the Second World War, suction dredgers were introduced and, with these, marine aggregates could be supplied at prices comparable with those from land sources. Initial doubts about the suitability of marine aggregate for concrete manufacture have been resolved and although sands are washed to remove alkalis, shells are no longer a problem in determining the suitability of much aggregate for construction purposes (Nunny and Chillingworth, 1986).

During a period of rapidly increasing offshore aggregate production in the UK during the 1960s, a new generation of dredgers was constructed averaging around 3000–4000 tonnes and being able to dredge to approximately 30–35 metres. However, some of the newer vessels have submersible dredge pumps located midway in the pipe, and can dredge to 45 m (Ardus and Harrison, 1989). In 1988

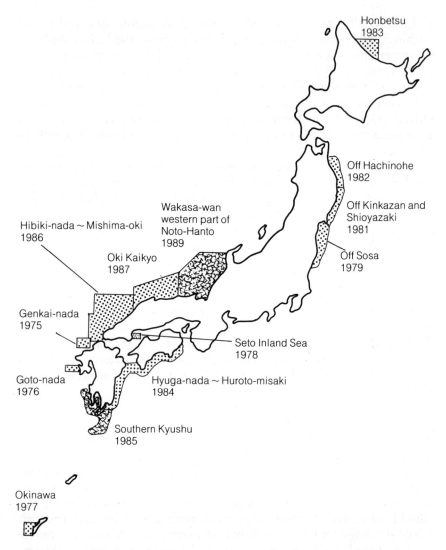

Fig. 2.3 Distribution of aggregate survey areas off Japan with dates of surveys.

there were about 50 dredgers working in UK waters, 45 of which were trailing dredgers (Fox, personal communication, 1988). High capital and operating costs, together with increasing capacity of the dredgers, have made it necessary to develop new techniques for discharging which speed up the operation. Most of the larger vessels today are equipped with self-discharging facilities. The ships need to

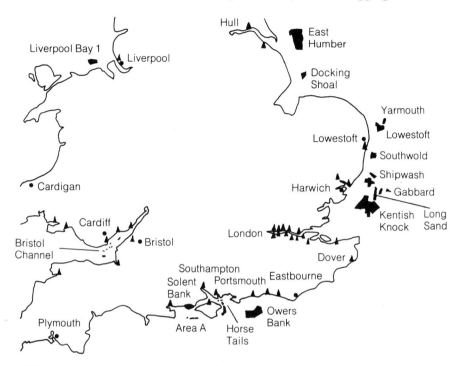

Fig. 2.4 Areas of aggregate dredging off the southern United Kingdom (from Uren, 1988). Triangles represent landing points.

run for 24 hours a day in order to recover capital costs. The unit value of fixed costs falls as throughput increases.

Over the past few years, as mentioned, the offshore aggregate industry has provided around 14–16% of the total UK consumption of sand and gravel. However, this percentage varies widely in different regions. In London and the South-East, for example, at times up to 43% of the area's total consumption has been contributed by marine dredged material (Webb, 1979), and on the south coast up to 50% (Parrish, 1988). Areas currently supplying aggregates to the UK market are shown in Figure 2.4.

Environmental and fishing objections have to be overcome in order to engage in offshore aggregate production but these can be placed in perspective when it is realized that only about 9 km² of seafloor is actually mined in any one year (Parrish, 1988). Additional production, if needed, could be obtained by dredging in water depths greater than the 35 metres usually dredged. In Japan aggregate dredging is commonly done down to 40–50 metres and this is starting to happen on the UK continental shelf. Dredging through overburden is another way in which offshore aggregate production could be increased. According to Webb (1979), there are substantial aggregate reserves which are covered by silt, or sand overburden, and which may be exploitable by using a technique which pierces through the overburden to reach the aggregate beneath. This has some environmental advantages in that it minimizes disturbance of the seafloor. However, it is likely to be a costly procedure and uneconomic at the present time. Webb (1979) also suggests that areas of contaminated aggregate might be brought into production, given that the necessary decontamination processes can be effected onshore.

3

Placers

Placers are metallic minerals or gems which have been transported to their sites of deposition in the form of solid particles, and are therefore detrital minerals. They are resistant minerals which have been made available on breakdown of their parent rock. Such rocks are usually igneous in origin, or veins, but sometimes the breakdown of metamorphic or sedimentary rocks can also liberate placer minerals into the secondary environment. All placer mineral deposits occur in EEZs.

3.1 NATURE OF PLACER MINERALS

The principal method of classifying placer minerals, other than by composition, is in terms of their specific gravity. Emery and Noakes (1968) devised a three part classification of placer minerals based largely but not entirely on specific gravities. They defined heavy heavy minerals as those with specific gravities of 6.8 to 21, light heavy minerals as those with specific gravities from 4.2 to 5.3, and gems with specific gravities from 2.9 to 4.1. Scheelite (SG 5.9–6.1) occupies an intermediate position. The heavy heavy minerals include noble metal deposits such as gold and platinum and the important tin bearing placer, cassiterite; light heavy minerals generally comprise the resistant accessory minerals of igneous rocks, and include monazite, zircon, ilmenite and rutile as important varieties, together with magnetite. The most important gem placer mineral is diamond. Properties of these minerals are shown in Table 3.1. According to Kudrass (1987), about 80% of total world rutile production, 50% of ilmenite and 30% of cassiterite were derived from marine sources in the mid-1980s. Most of the world's zircon is also of placer origin.

3.2 FORMATION OF PLACER MINERAL DEPOSITS

The formation of placer mineral deposits requires several separate processes. Naturally there has to be a primary source of minerals,

Table 3.1. Principal placer minerals and their composition

Gems	Specific gravity	Composition and principal element
Diamond	3.5	C
Garnet	3.5–4.27	$(CaMgFeMn)_3 (FeAlCrTi)_2 (SiO_4)_3$
Ruby	3.9–4.1	Al_2O_3
Emerald	3.9–4.1	Al_2O_3
Topaz	3.4–3.6	$Al_2F_2SiO_4$
Heavy noble metals		
Gold	20	Au
Platinum	21.5	Pt
Light heavy minerals		
Beryl	2.75–2.8	$Be_3Al_2Si_6O_{18}$ (Be)
Corundum	3.9–4.1	Al_2O_3 (Al)
Rutile	4.2	TiO_2 (Ti)
Zircon	4.7	$ZrSiO_4$ (Zr)
Chromite	4.5–4.8	$FeCr_2O_4$ (Cr)
Ilmenite	4.5–5.0	$FeOTiO_2$ (Ti)
Magnetite	5.18	Fe_3O_4 (Fe)
Monazite	5.27	$(CeLaYt) PO_4 ThO_2 SiO_2$ (Th REE)
Scheelite	5.9–6.1	$CaWO_4$ (W)
Heavy heavy minerals		
Cassiterite	6.8–7.1	SnO_2 (Sn)
Columbite, Tantalite	5.2–7.9	$(FeMn) (NbTa)_2O_6$ (Nb, Ta)
Cinnabar	8–10	HgS (Hg)

usually igneous or metamorphic rocks, the resistant accessory minerals which, as mentioned, are common constituents of light heavy mineral placer deposits. These minerals are liberated on weathering of their source rocks, and are released for transport. Chemical weathering is more important than mechanical weathering in this regard. This is one of the reasons why placer mineral deposits are more common in mid- and low latitude areas than they are in high latitude areas where mechanical weathering is predominant. Transport of the liberated placer minerals is largely by streams and rivers, during which the unstable products of the weathering such as clay minerals are separated from the heavier more resistant placers. Having finally reached their site of deposition, either beaches or near shore environments, marine placers are usually concentrated in sinks or traps because of their high specific gravity, or on the upper parts of beaches. During this process lighter and smaller minerals become separated from larger or heavier minerals, leading to a concentration of the placer minerals to potentially economic grades.

The depositional environments of marine placer minerals have been reviewed by Cronan (1980) and need not be considered in detail here. It should be remembered, however, that some of the heavy heavy placer mineral deposits currently offshore were formed in river valleys which subsequently became inundated during the post glacial rise in sea level. Certain deposits of cassiterite, scheelite, gold and platinum fall into this category. Further, many placer minerals are deposited on beaches near the mouths of the rivers down which the minerals were transported. For the purposes of this work, both these types of deposits are classified as marine placers.

Placer mineral formation on beaches is essentially the result of selective sorting of the beach deposits in the inter-tidal zone by wave and current action, often in longshore drift. Sorting of heavy minerals and separation of different placer mineral varieties is mainly due to selective mobilization and transport of minerals according to their grain sizes and densities. The smaller the grain size (partly due to entrapment) and the higher the density, the greater is the chance that the mineral will not be picked up by the backwash once deposited by the swash (Komar and Wang, 1984). Placer minerals are transported up the beach in the swash where the backwash is only strong enough to remove lighter minerals, and thus they concentrate generally on the upper portions of the beach. In many parts of the world, black sand deposits formed by this process can be seen in the middle and upper reaches of beaches. Beach placer minerals generally comprise the light heavy minerals of Emery and Noakes (1968) as the specific gravity of these is such as to enable them to be moved about during beach processes. However, processes forming beach placer mineral deposits are not stable. They can be affected by seasonal and latitudinal variations in the strength of the waves, and in high latitudes beaches can vary widely in extent from one season to another due to storm activity. Beach mean grain size can also vary from season to season, and this can have a marked effect on the presence or absence of heavy mineral deposits. The best situation for the formation of a light heavy mineral placer deposit on beaches is one in which the waves are relatively constant, both in strength and direction, throughout most of the year. High energy low- to mid-latitude beaches are thus likely to be the most favourable environments for light heavy mineral concentration to take place, assuming, of course, that there is a suitable source of minerals in the hinterland, and there is no reef offshore to dissipate the wave energy.

Many beaches are characterized by having a longshore trough between the breaker zone and the swash zone. In this can sometimes occur gravel deposits in which small dense heavy heavy minerals such as gold can become trapped. In the absence of the gravel, these

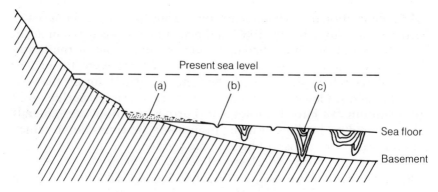

Fig. 3.1 Environments of offshore placer mineral occurrence (modified from Cronan, 1980, after Aleva, 1973): (a) placers on submerged beaches; (b) placers trapped in surface depressions on the sea floor; and (c) placers in buried river valleys.

would continue to be transported, but once trapped between the gravel particles they are protected from further transport.

In the offshore area, placer minerals can occur in several situations (Figure 3.1). Submerged beaches are one such situation. The deposits would have been laid down on a normal beach during a low stand of sea level, and subsequently submerged when sea level rose. Many continental shelves contain submerged terraces which were once beaches and which are usually parallel to the present coastline. However, such beaches generally contain fewer placer minerals than their modern equivalents because during post-glacial transgressive sea level rises, the placer mineral deposits on the beach are either disseminated seawards or tend to follow the landward moving surf zone (see Mozambique and Australian case studies, pp. 45 and 48). Submerged beaches are therefore generally less attractive than modern beaches as far as placer mineral deposits are concerned.

As mentioned, drowned river valleys can contain fossil alluvial placer mineral deposits as, for example, off south-east Asia. Often, however, in these situations, the placer minerals are at the bottom of the sediment column, overlying bedrock, rather than in a more exploitable situation near the surface (Cronan, 1980).

Placer minerals transported from the land into the offshore area can be further transported by wave and current action, but when the energy level of the environment decreases to below that at which the minerals can be moved around, they are deposited on the sea floor. Once deposited, a significantly higher energy level is required to pick them up again and in this way localized concentrations of placer minerals can occur. Hosking and Ong (1963) noted, for example, that off north Cornwall in the UK, free cassiterite grains tend to be concentrated in elongate zones below low-water and parallel to the coast. An extreme example of this sort of process has been reported by Moore and Welkie (1976) who have shown that very fine grained gold and platinum placer deposits (5–50 μm) can occur in quiet bays and protected estuaries off Alaska. Initially supplied to the offshore area under higher energy conditions, their distribution is believed to be related to circulation gyres of slackening currents within protected embayments, coupled with agglomeration of particles on contact with electrolytes in seawater.

3.3 PLACER MINERAL OCCURRENCES

Major placer mineral occurrences around the world are shown in Figures 3.2–3.5.

3.3.1 Gems (Figure 3.2)

The principal gem placer minerals occurring in the offshore environment within Exclusive Economic Zones are diamonds off South-West Africa. These were first discovered in raised beach sands in 1908 (Murray, 1969), the diamond bearing beds extending down to the coast and offshore. The diamonds are derived from the interior and have been transported to the coast by fluvial processes. They have been mined offshore intermittently for several years since the early 1960s, and more recently by systematic traverse suction dredging. Quantitative extraction of the overburden is required as the diamonds are often resting on bedrock, or in cracks and joints in it. The deposits have been recovered by using beach sand overburden to build temporary paddocks to hold back the sea and thus allow onshore working techniques to be used below the sea level.

Diamonds are also thought to occur off Venezuela, India, China and northern Australia, and garnets have been reported on beaches off Fiordland, New Zealand (Summerhayes, 1967).

3.3.2 Heavy heavy minerals (Figure 3.3)

The most common of the heavy heavy placer minerals are the noble metals, gold and platinum, and the tin mineral, cassiterite (Table 3.1). Placer gold has been sought offshore of many gold mining areas, and has been recovered from several coastal and nearshore environments. At Paracale Bay in the Phillipines, one of the largest

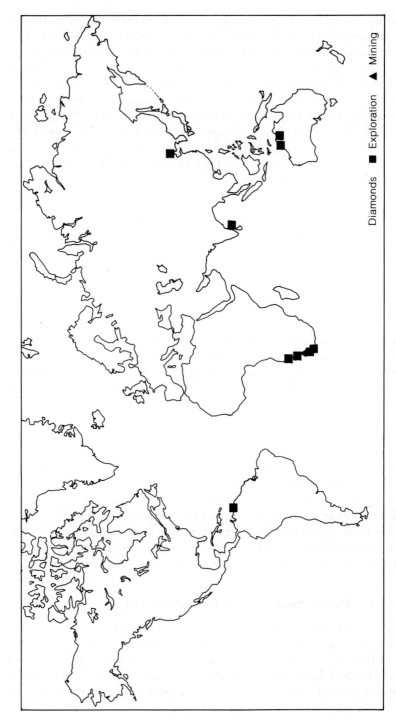

Fig. 3.2 World distribution of offshore diamond occurrences (after Hale, 1988).

Diamonds   ■ Exploration   ▲ Mining

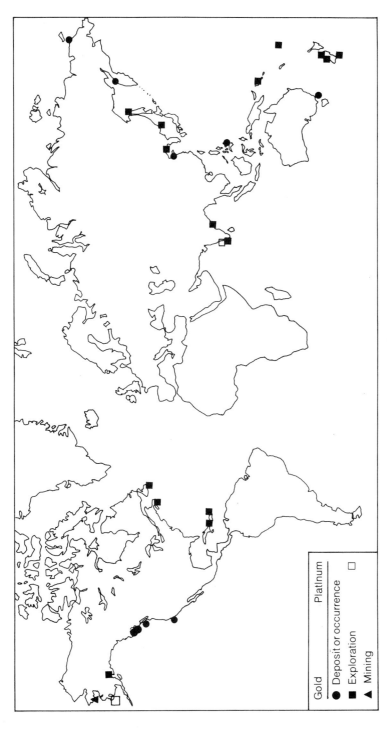

Fig. 3.3 World distribution of noble metal placers (after Hale, 1988).

and most successful offshore mining operations for gold produced more than 15.5 million grams (500 000 ounces) by 1922. Placer gold has been mined on the beaches and more recently offshore of Alaska, principally in the district around Nome. There has been a resurgence of interest in the gold placer mining industry in recent years as the gold price has increased, and this has led to a number of new developments being opened or old developments being re-examined.

During the early 1960s, in Nova Scotia, more than 62 000 grams (2000 ounces), of gold were recovered from beach sands at the Ovens. An extensive programme of offshore exploration subsequently led to the discovery of two areas along the coast of Nova Scotia that were estimated to have combined volumes of 32 million cubic metres of gold bearing sediments. On the other side of Canada, off British Columbia, more than 68 000 grams of gold have been recovered along beaches on Queen Charlotte and Vancouver Islands. Small deposits of gold have been reported off New Zealand (Summerhayes, 1967).

One of the most recent gold placer mining operations is, as mentioned, off Nome, Alaska. This is an area that has yielded gold both onshore and on the beaches, about 4 million ounces in all, since 1898. The high concentrations of gold that have been panned from the beaches led to speculation for a long time that further significant gold deposits might occur offshore. Various methods were used in an attempt to locate and mine such gold but until recently none were particularly successful. The most successful effort involved the use of an Indonesian tin dredger, the *Bima*, brought from south east Asia in 1986 when it was no longer needed for cassiterite mining. According to Garnett (Underwater Mining Institute, oral communication, 1989) about 100 000 ounces of gold were extracted over the three seasons 1987–9 using this dredger, all from within the area three miles seawards and thus under State jurisdiction.

According to Hale (1988), occurrences of gold have been reported off a number of other countries (Figure 3.3).

Although gold is the principal noble metal placer mineral currently being exploited within EEZs, it is not the only one of interest. Platinum has been found to occur off Alaska in the form of very fine grained particles, the source probably being nearby ultramafic rocks (Moore and Welkie, 1976). It has also been reported off India (Hale, 1988) (Figure 3.3).

The principal non-noble metal heavy heavy mineral placer is cassiterite, tin oxide, and is probably the offshore placer mineral that has been mined in greatest abundance. South-east Asia hosts the world's largest offshore cassiterite deposits, off Indonesia, Malaysia and Thailand. These deposits fall mostly into the category of placers which were formed in river valleys which have now become submerged below sea level as a result of the post-glacial rise in sea level. Indeed, according to Hosking (1971) the post-glacial rise in sea level can be held responsible not only for the submergence of the deposit, but also at least in part for its origin. He concludes that major fluctuations in sea level during Pleistocene times played a major part in inducing rejuvenation of the rivers in the tin-bearing area leading to rapid valley erosion, enhanced transport of the placers by rivers, and their subsequent re-working when sea level rose.

Placer tin deposits are widely distributed off south-east Asia and have been reviewed by Hosking (1971) (Figure 3.4). Cassiterite deposits have been recovered for a considerable time adjacent to Phuket Island off Thailand and the tin islands off Indonesia. In Malaysia, the western tin belt close to the west coast of the Malay peninsula is richer in cassiterite than the eastern tin belt, and cassiterite has been recovered from beach sands and other coastal deposits on the west coast. Between Thailand and Burma, estuarine alluvium contains cassiterite which might be of future economic value. Indeed, according to Hosking (1971), much of offshore Burma has a tin potential.

The mechanism of formation of the south-east Asian tin deposits can largely be related to processes operating during Pleistocene times. According to Hosking (1971), cassiterite accumulated on hillsides and valley slopes as a result of weathering of primary deposits and some of it migrated under gravity to form deposits at the base of the slopes. Some also was transported into rivers where it tended to become concentrated by the preferential removal of minerals of lower specific gravity. Because of the brittle nature of the cassiterite, grains were reduced to small particles which could be transported further than would otherwise be the case. However, Hosking (1971) believed it unlikely that economically important deposits of alluvial cassiterite would occur very far from their primary source and suggests that the median distance from the source to the deposits is perhaps only about 8 km. Cassiterite from sources near the coast was transported by rivers to the sea and some of the finer fraction was transported by longshore currents. Marine transgression caused the reworking and upgrading of some of the alluvial deposits and the local deposition of low grade tin bearing sediments. As erosion slowed down with regrading of the rivers, the amount of cassiterite added to the alluvium decreased so that barren

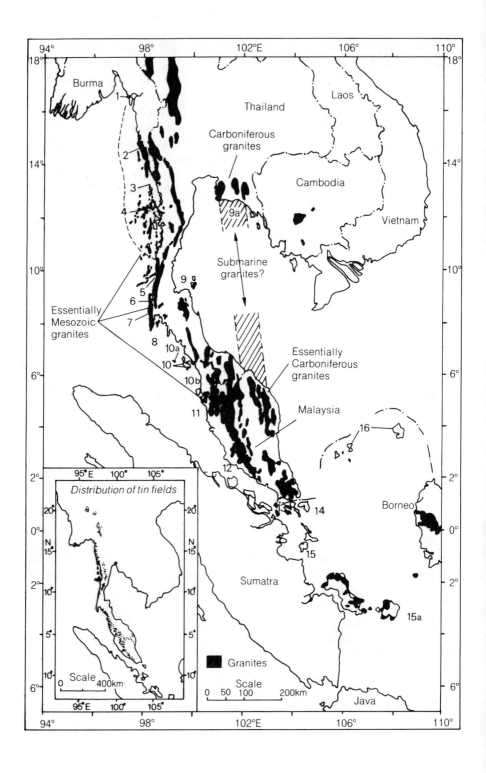

Burma

1

2

3

4

Thailand

Carboniferous
granites

Laos

Cambodia

Vietnam

9a

Submarine
granites?

Essentially
Carboniferous
granites

Essentially
Mesozoic
granites

9

5

6

7

8

10a

10

10b

11

12

Malaysia

16

14

Borneo

15

Sumatra

15a

Java

Granites

Scale
0 50 100   200km

Distribution of tin fields

Scale
0   400km

Fig. 3.4 The offshore tin areas of south-east Asia, and the general distribution of granites and tin belts (after Hosking, 1971). Locations 1–16 are as follows:

1. Belugyun Island: Stanniferous placers.

2. Heinze Basin: Cassiterite dredged from tideways.

3. Spider Island (at mouth of Palauk River): Cassiterite (and wolframite) recovered from beach sands.

4. Tenasserim Delta and Lampa and neighbouring Islands: Cassiterite recovered from these localities.

5. Ranong and coast to South: Onshore placers locally extend to coast. Primary tungsten deposit at coast (at Kau Chai).

6. Takuopa: Suction dredge has worked offshore.

7. Thai Muang: Cassiterite in beach sands.

8. Phuket. Some bays on west coast known to contain important concentrations of cassiterite. Past illicit offshore mining off west coast.

9. Ko Phangan and Ko Samui: Poor tin mineralization on islands. Cassiterite occurs offshore.

9a. Rayong: Cassiterite in beach sands and offshore.

10. Langkawi Islands: 2% of Sn in andradites; Stannite (slight) locally present. "Trace" of cassiterite in beach sands.

10a. Ko Ra Wi and Ko La Dang (WNW of Langkawi): Cassiterite reported.

10b. Islands west of Kedah Peak: Cassiterite in beach sands.

11. Lumut-Dindings: Cassiterite in beach sands and offshore at Tanjong Hantu and Pulau Katak. Site of past illicit offshore mining.

12. Malacca: Cassiterite in mainland beach sands and offshore, from S. Linggi to S. Udang, also in beach sands of Pulau Besar (an island).

13. Karimun and Kundur: Tin-bearing islands but mineralization not strong.

14. Bintan: Small placers on island.

15. The Tin Islands: Substantial past offshore mining and exploration activity.

15a. Submarine tin granites between Billiton and Borneo.

16. Anambas and Natuna Islands: Cassiterite reported but outside tin belt proper.

alluvium commonly overlies the tin placers. In essence then, the offshore placers of south-east Asia are principally non-marine in origin, but have become submerged and reworked as a result of transgressive sea level rise invading the valleys in which they were deposited.

Other occurrences of offshore tin have been reported off North America, Europe, Australia and the USSR (Hale, 1988).

3.3.3 Light heavy minerals (Figure 3.5)

Light heavy minerals (Table 3.1) have the widest distribution of all classes of placer minerals (Figure 3.5). As mentioned earlier, they occur principally on beaches, modern or drowned, and in the immediate offshore area.

Off North America, light heavy minerals have been found off Florida and South Carolina, and there is considerable interest in heavy mineral exploitation along the south-eastern Atlantic Seaboard of the United States (Richey, 1988). Heavy mineral concentrations coupled with magnetic anomalies on the southern Oregon continental shelf indicate that placer minerals may occur at shallow depth below the offshore sediments there (Kulm, 1988). These could include chromite, ilmenite and zircon; and indeed chromite and ilmenite were mined on coastal terraces of south-west Oregon during the Second World War (Griggs, 1945). Gold has also been reported (Richey, 1988). Other coastal light heavy placer mineral deposits elsewhere in the US Pacific north-west include ilmenite and magnetite (Peterson and Binney, 1988).

Off Europe, there has been little mining of placer mineral deposits. Cassiterite deposits off Cornwall will be reviewed shortly. A small magnetite deposit has been reported off Norway at the latitude of the Lofoten Islands (Sandvik, personal communication, 1982), but has not been mined. Varnavas (1986) has reported occurrences of light heavy minerals off Cyprus in the eastern Mediterranean Sea, and Perissoratis et al., (1987, 1988) in the northern Aegean Sea. Chromite deposits occur off the Isle of Rhum, Scotland (Ardus, personal communication, 1986).

South-east Asia contains some important light heavy mineral deposits, particularly off India and Sri Lanka. In Sri Lanka, rutile and zircon have been produced from the beaches, and these minerals, together with others have a widespread distribution on both the east and west coasts of peninsular India. According to Siddiquie and Rajamanickam (1979), major placer deposits occur on the coast of Kerala (India) and contain ilmenite, rutile, zircon and monazite.

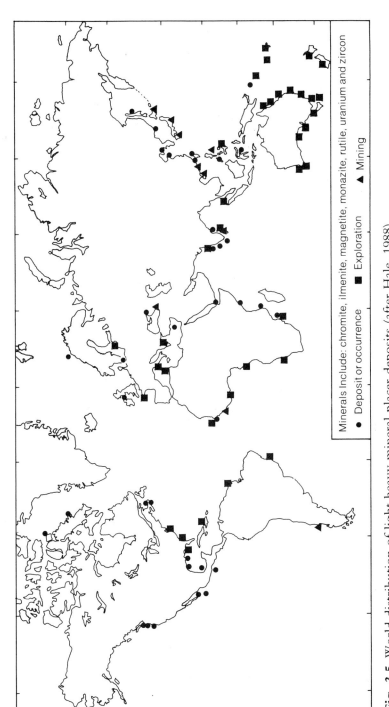

Fig. 3.5 World distribution of light heavy mineral placer deposits (after Hale, 1988).

Minerals Include: chromite, ilmenite, magnetite, monazite, rutile, uranium and zircon

● Deposit or occurrence　■ Exploration　▲ Mining

Similar beach deposits, also containing magnetite and garnet, have been found on the Maharashtra (India) coast further north, and on the east coast of India. In 1979 the total resources of ilmenite were estimated to be 138 million tonnes. Placer mineral deposits also occur offshore, within the Indian EEZ. Prabhakar Rao (1968), for example, reported from 0.8–20% ilmenite in sands off the Kerala coast, and Siddiquie *et al.* (1979) report up to 90% heavy minerals, mainly ilmenite and magnetite, in offshore sediments off the Konkan Coast of India. Magnetite deposits used to be mined off Japan, as was garnet, but this has now ceased (Arita, personal communication, 1989).

Light heavy minerals have been reported at various locations around Africa (Figure 3.5) (Cronan, 1980). Most of these are off sub-Sahara Africa and are probably related to the large rivers that drain the interior of the Continent. Deposits of rutile, ilmenite and zircon off Mozambique are the subject of a case study on page 45. Rutile also occurs off Sierra Leone and ilmenite off Senegal.

Australian light heavy mineral mining has been of considerable importance for some time. Monazite, rutile and zircon sands occur on beaches in Queensland and in New South Wales (Dunham, 1969), and vary from a few inches to several feet in thickness. Dunham (1969) reported that around Bunbury in Western Australia beds of heavy minerals up to 7.5 m thick, principally of ilmenite, occur at present sea levels and higher. Rutile and zircon bearing sands have also been found off East Australia and in the area of south-east Australia and Tasmania (Emery and Noakes, 1968; Jones and Davies, 1979). Light heavy minerals off south-eastern Australia are the subject of a case study on page 48.

Light heavy minerals have also been reported off the coast of New Zealand (Summerhayes, 1967). Titanomagnetite occurs on raised beaches on the west coast of the North Island. Ilmenite beach deposits occur near Auckland and also on the west coast of South Island where small deposits of magnetite, rutile, monazite and zircon also occur. Iron sand concentrations have been found in 27 m water depth off North Island. Additional occurrences of light heavy placer minerals off New Zealand have been reviewed by Glasby (1986).

3.4 CASE HISTORIES

In order to illustrate the diverse environments and conditions under which placer minerals can form, four examples have been chosen for more detailed treatment. There are:

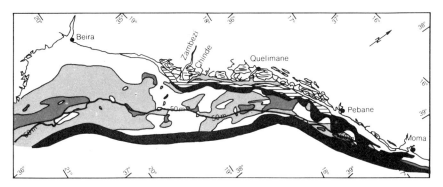

Fig. 3.6 Sediment types offshore and locations of coastal beach ridges, Mozambique (from Beiersdorf *et al.*, 1980).▤ Mud with <10% sand,█ mud with 10–50% sand,☐ fine sand (63–250μ),▨ medium sand (250–500μ),▧ coarse sand (>500μ),▧ beach ridges.

1. a large disseminated placer in a continental margin setting off Mozambique;
2. concentrated placers on south-eastern Australian beaches;
3. a man-made placer off Cornwall; and
4. placers associated with south-west Pacific islands.

3.4.1 Mozambique continental margin

The Mozambique continental margin (Figure 3.6) is a passive one, along the boundary of which have accumulated large masses of sediment largely derived from the erosion of crystalline rocks in the hinterland. The heavy minerals which comprise the placer deposits off Mozambique are the accessory minerals of these crystalline rocks, and are estimated to total 50 million tonnes of ilmenite, 4 million tonnes of zircon and 0.9 million tonnes of rutile (Biersdorf *et al.*, 1980).

There are several known placer deposits onshore in coastal Mozambique containing ilmenite with considerable amounts of zircon and rutile (Biersdorf *et al.*, 1980 and references therein). Deposits at Pebane (Figure 3.6) were mined in 1959 and 1960. Other deposits immediately north of the mouth of the Zambezi River also contain ilmenite, zircon and rutile. Approximately 50 million tonnes of heavy mineral rich sand has been estimated to occur in this deposit. These placers were formed by wave, current and wind action along the modern and recent shorelines. The occurrence of

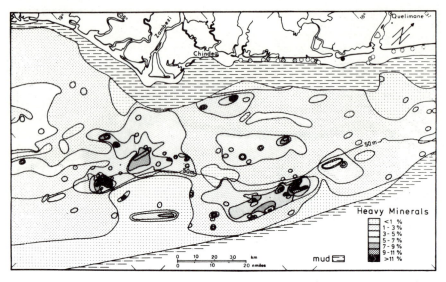

Fig. 3.7 Distribution of heavy minerals in the fine sand fraction of sediments off the Zambezi Delta (from Beiersdorf *et al.*, 1980).

these placer deposits onshore in Mozambique suggested that similar deposits might have been formed during the Pleistocene period in areas that were then dry land but which have subsequently been submerged by the post-glacial rise in sea level. This possibility provided the impetus for a comprehensive offshore placer mineral survey reported on by Biersdorf *et al.* (1980).

In order to evaluate the placer potential of the Mozambique shelf, more than 1000 surface and core samples were collected on the continental margin and were subjected to detailed sedimentological and geochemical studies. This proved the occurrence of a large ilmenite-zircon placer sand off the Zambezi Delta in water depths between 30 and 60 metres. This more or less coincided with a 3 to 10 metre high step along the 50–55 metre isobath which was thought to represent an ancient shoreline (Figure 3.7).

The heavy minerals in the shelf sediments can be divided into opaque and transparent varieties. The former include ilmenite, leucoxene, magnetite, titanomagnetite and spinels, and the latter include titanite, pyroxene, epidote, garnet, zircon, tourmaline, rutile, kyanite, sillimanite and several others of minor importance. The areas with more than 3% heavy minerals are concentrated between the Zambezi Delta and Quelimane and within this zone there are four sub-areas where average heavy mineral concentrations exceed 5% (Figure 3.7).

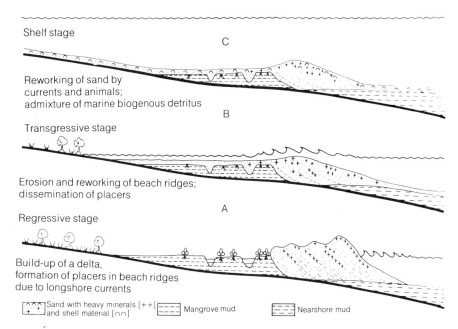

Fig. 3.8 Suggested origin of offshore Mozambique placer deposits (from Beiersdorf *et al.*, 1980).

At the present time, there is no significant transport of sand from rivers to the outer continental shelf, since most of the sand is trapped in a longshore transport system. The heavy mineral bearing non-biogenic fraction of the shelf sediments must therefore have been laid down during the Pleistocene period when the shelf was dry during a period of lowered sea level. The lowering of the sea level was interrupted by a stillstand phase which formed a shoreline marked by the scarp at about 50–55 m depth. The formation of heavy mineral bearing beach ridges in the vicinity probably took place under conditions of longshore drift similar to those forming such ridges at the present time in the Zambezi Delta.

As the sea level started to rise again around 15 000 years ago, the transgressing sea partially reworked and redeposited the Pleistocene shelf sediments (Figure 3.8). This resulted in an intensive destruction of the heavy mineral stringers in the ancient beach ridge system. However, the heavy mineral content was kept more or less in place, rather than being disseminated seawards or transported landwards, as is indicated from the occurrence of the area of high heavy mineral concentrations in the immediate vicinity of the old shore line.

To summarize the origin of the Mozambique heavy mineral sands (Figure 3.8), there was a stillstand during a general lowering of sea level which allowed the formation of an extensive beach ridge system containing heavy mineral stringers at an ancient Zambezi Delta. On rising sea level, this beach system was partially destroyed by the marine transgression and subsequently reworked by currents. This reworking led to the homogenization of the beach ridge sands and heavy mineral stringers, thereby disseminating the heavy minerals throughout a larger volume of sand than was the case when they were formed. The location of the heavy mineral enrichments remains the same, however, but the heavy minerals were diluted in situ by mixing with their associated barren sands. The sand body off the modern Zambezi Delta still has a high heavy mineral content and may be workable. It provides a good example of a large placer deposit formed off a major land mass containing abundant heavy mineral bearing rocks.

3.4.2 South-eastern Australian continental shelf

Unlike the Mozambique shelf, the eastern Australian continental shelf is narrow, (10–20 km) with a shelf break at 70–150 m (Figure 3.9). Distributed along the coast are Holocene and Pleistocene beach placer deposits which have been mined for rutile, zircon, monazite and ilmenite. However, many of these deposits are exhausted or near exhausted and this prompted the evaluation of the offshore area for similar mineral deposits (Jones *et al.*, 1982).

Placer mineral exploration investigations were carried out in four shelf areas (Figure 3.9). In most of these areas, the Holocene sand of the outer and middle shelf is only a few metres thick. Potential sites for Pleistocene beach placer deposits are in barrier systems which occur offshore. However, it was found that the Holocene beach sand contains up to ten times greater concentrations of heavy minerals than in the Pleistocene sand on the shelf. Generally there are differences in the heavy mineral assemblage of the different sands, the Holocene beach sands being relatively rich in potentially economic heavy minerals like rutile, tourmaline, zircon and ilmenite, whereas the Pleistocene shelf sands contain predominantly uneconomic heavy minerals like amphibole, epidote and pyroxene.

The differences between the Holocene beach and Pleistocene shelf sands are explicable in terms of processes that took place during the post-glacial transgressive sea level rise (Figure 3.10) (Kudrass, 1987). The transgressive Holocene shelf sand provides a link between the Holocene beach and Pleistocene shelf sands which can help to explain the differences between them. In the Holocene shelf sand, there is a

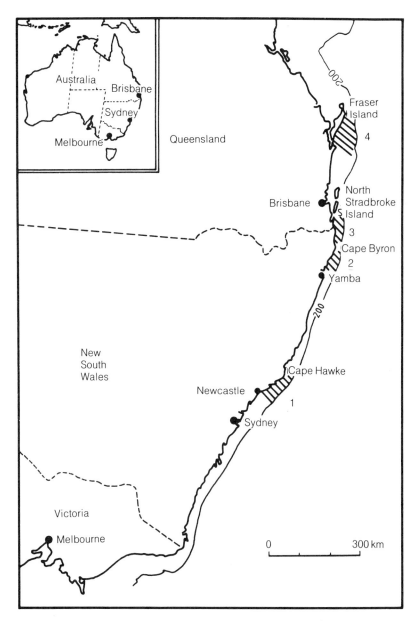

Fig. 3.9 Placer mineral investigation areas off Australia (from Jones *et al.*, 1982).

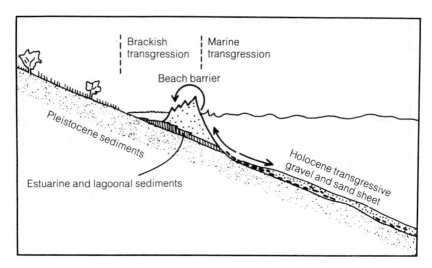

Fig. 3.10 Development of early Holocene beach barriers and origin of the transgressive sand and gravel sheet (from Kudrass, 1987). The beach barrier moved landwards during the rapid rise of sea level.

correlation of heavy mineral composition with water depth which can be explained by a combination of two processes; (a) beach barrier migration during the Holocene transgression; and (b) vertical sorting of heavy minerals in beach sand. Beach sorting processes result in vertical differentiation of heavy minerals within the beach sand on the basis of their densities. The heavier minerals such as magnetite, ilmenite, zircon and rutile, are found enriched in the upper beach sediments whereas the less dense minerals, such as epidote and pyroxene are enriched in the lower and offshore parts of the beach. For example, the nearshore sand contains 9% epidote, whereas the sand above the shoreline contains only 1%. During beach barrier migration, a sand sheet was left behind on the transgressed shelf containing reworked Pleistocene sediments at its base. During the transgression, the lighter minerals were removed from the heavier minerals in the beach barrier sand by density sorting and preferentially deposited in the nearshore sand. By contrast, the more economically important minerals, rutile, zircon and ilmenite, were preferentially transported landwards as the beach migrated landwards and their concentration in the beach sand continuously increased (Figure 3.10). So, while the sand being swept up on to the coastline was only a small proportion of the total volume of sand moved, it contained a much higher proportion of heavy minerals, which provided the beach placer deposits which have been mined along the east Australian coast.

3.4.3 Comparison of Mozambique and south-east Australian placer mineral deposition

The differences between the placer mineral deposits off Mozambique and south-east Australia provide important information not only on mechanisms of formation of placer mineral deposits, but also on situations in which different types of deposits might be sought.

In the Mozambique situation beach placer deposits were formed in the outer delta of a large river on what is now the continental shelf, during the Pleistocene lowering of sea level. The shelf is wide, terrigenous supply was moderate to high, and the coast was a high energy one. During the post-glacial rise in sea level, the beach ridges containing placer deposits were reworked and became homogenized in the vicinity of the original coastline. Possibly because of the broad, gently sloping shelf, there was insufficient energy to move the large amounts of beach sand landwards as the sea transgressed and therefore it became redistributed more or less in the area in which it was deposited. The gently sloping nature of the shelf would mean that for every given incremental rise in sea level, a large horizontal distance would have to be traversed, and the energy of the waves was insufficient to transport the sand over these distances.

Off south-east Australia, the situation is different in that the continental shelf is narrow, and the amount of material supplied to the shelf was low to moderate in abundance. During beach formation, vertical sorting of heavy minerals took place with the more economically valuable minerals enriched in the upper parts of the beach profile and the lighter heavy minerals being concentrated in the lower part of the beach profile. During the transgression associated with the post-glacial rise in sea level, the beach barriers migrated landwards and during this migration the upper parts of the barriers were preferentially transported shorewards leading to the concentration of economically valuable minerals increasing continuously as landward migration continued. In this situation placer minerals which were originally present on the shelf have essentially been swept up on to the shore, and thus the likelihood of there being similarly rich placer deposits offshore is low.

One may conclude from these studies that where there are rich beach placer deposits onshore, the heavy mineral potential of the shelf area is likely to be low in comparison with those beaches. By contrast, broad shelves where the rate of terrigenous supply is high

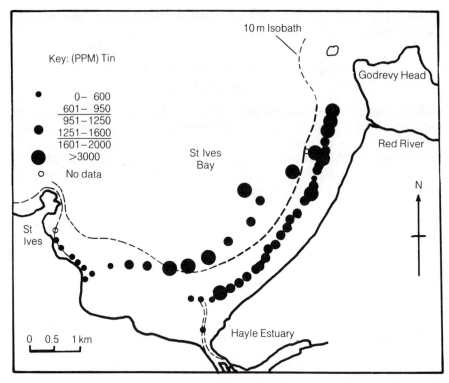

Fig. 3.11 Distribution of tin in the sand fraction of sediments from St Ives Bay (from Yim, 1979).

may have a good potential for hosting disseminated placer deposits but these are likely to be lower grade than present day beach placers along their coasts.

3.4.4 Cassiterite deposits off Cornwall (UK)

During 1961–5 a Union Corporation subsidiary company sought for offshore tin placers in St Ives Bay, Cornwall (Figure 3.11). An area in about 10 m of water averaging 0.2% tin was discovered and mined for a period.

The hinterland of St Ives Bay consists essentially of shales and sandstones intruded by granite. Associated with the latter, are tin-bearing mineral lodes which have been worked intermittently over a very long period of time. This has resulted in the contamination of soils and stream sediments by mine tailings. Rich tin slimes from the South Crofty and Pendarves tin mines have been transported down the Red River into St Ives Bay (Yim, 1979). This transport is likely to have been the main cause of the enrichment of placer tin in St Ives Bay.

The richest tin area in St Ives Bay occurs off the mouth of the Red River with other areas of tin enrichment in sediments off the Hayle Estuary and in the middle portion of St Ives Bay (Figure 3.11). Overall, the average tin content in the sand fraction of the superficial sediment is about 0.1% tin oxide.

The tin enrichment off the Hayle Estuary can be explained by the sluicing of alluvial sediment containing mine tailings out of the estuary. Likewise, the enrichment off the Red River can be explained by mine tailings from the Camborne–Redruth area. However the highest tin values off the Red River occur close to Godrevy Head, indicating that modern marine wave and tidal processes must be redistributing the tin supplied by the Red River and selectively concentrating it. The high tin values in the middle of St Ives Bay, off the Hayle Estuary, are also probably due to modern marine concentration processes, most probably bottom current action. These high values are superimposed on a general decrease in the tin content of the sediments away from the shore.

A mining contamination origin for the tin in the sediments of St Ives Bay is supported by drill hole tests at the mouth of the Red River. Tin enrichment only occurs in the uppermost 0.6 m of sediment.

3.4.5 Placer minerals in the tropical south-west Pacific

Unlike the other case studies on placers considered in this work which are in continental margin settings, this one deals with small occurrences in tropical island settings.

Some of the largest placer mineral deposits in the south-west Pacific occur off Vanuatu. Vanuatu consists of a number of islands distributed along an approximately north-south chain centred on 16°S, 168′E (Figure 3.12). The islands are composed predominantly of basic volcanic rocks.

Because of their young topography and predominantly wet climate, river action on the islands is considerable, leading to rapid erosion of the volcanic rocks. This river action liberates magnetite and other minerals from the rocks and transports them down to the

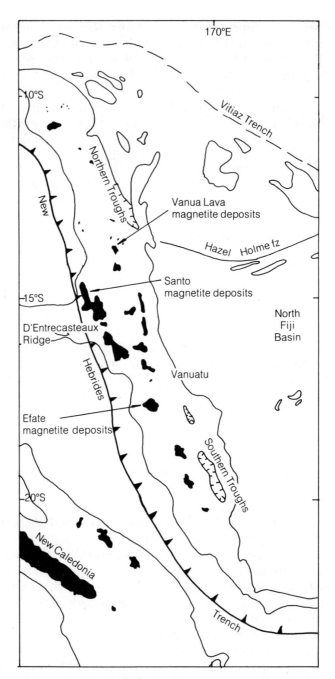

Fig. 3.12 Placer mineral bearing islands of Vanuatu (from Cronan, 1983) and locations of hydrothermally active troughs.

sea. This process leads to the development of magnetite deposits in some coastal areas, the sizes of the deposits being related to the sizes of the source areas and the sizes of the rivers eroding them (Exon, 1981a). The grades of the deposits in weight per cent magnetite varies dependent upon the degree of reworking they have undergone. Where the offshore area is fringed by a reef, the grade of the magnetite sand is low, because wave action is reduced by the reef.

Of the magnetite deposits known in Vanuatu, most of the largest deposits occur on Santo, the largest island. Deposits elsewhere are smaller. This is largely a reflection of the size of the source areas. However, the deposits on Santo are for the most part low grade. On Efate, a deposit estimated to yield 173 000 tonnes of magnetite has been reported. However, these deposits are rich in titanium which mitigates against their use as a source of iron ore.

One well investigated magnetite deposit occurs at Port Patteson (Figure 3.13) on the island of Vanua Lava where beach sands containing up to 22% magnetite occur, with pyroxene and olivine comprising the bulk of the remaining minerals (Exon, 1981b). Magnetite enrichment also occurs offshore at Port Patteson. This is due to the deposition of fine grained magnetite which has by-passed entrapment on the beaches.

The Vanuatu magnetite deposits are currently uneconomic, but provide a good example of small placer deposits formed in an ocean island setting. In such an environment, placer deposits are never likely to be very large because the source areas from which the minerals are derived are not large. Nevertheless, they may be economically important if they are of high value or can satisfy a local need.

The Solomon Islands are another area where placer mineral deposits occur (Biliki, 1988). Minerals include ilmenite, magnetite, zircon, gold, chromite, hornblende and pyroxene (Figure 3.14). Dominant are ilmenite, chromite and magnetite. All three occur on Choiseul, Santa Isabel, Malaita and Guadalcanal. A sub-economic olivine beach sand has been reported from Baniata Point, Rendova Island and some olivine beach sands are known to occur at Ranongga. All these beach deposits are Quaternary in age and originate from andesite, basalt or ultrabasic rock formations inland (Biliki, 1988). However, only a few reserve estimates have been made. There are 2 million tonnes of ferromagnesian sand at Baniata Point, Rendova Island and 9228 tonnes of magnetite has been found at Lithogohire Bay, West Santa Isabel. Some potential chromite

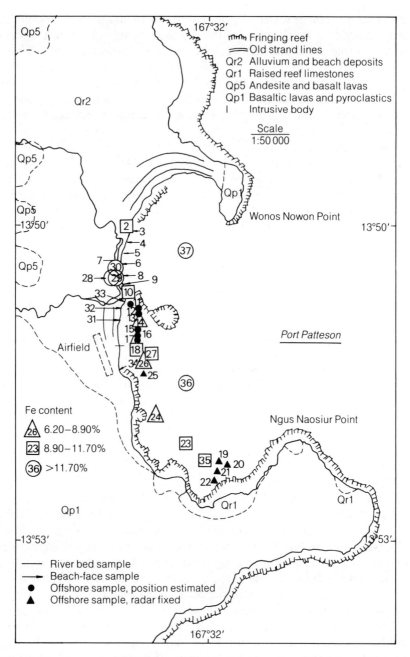

Fig. 3.13 Magnetite in deposits off Port Patteson (modified from Cronan, 1983).

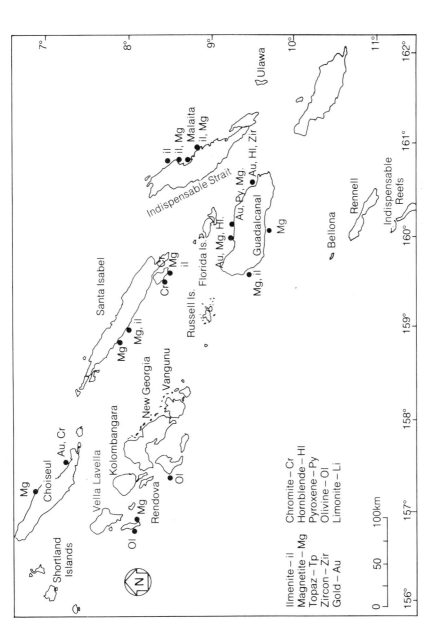

Fig. 3.14 Nearshore placer mineral occurrences in the Solomon Islands (after Biliki, 1988).

deposits occur on the beaches of east, north-west and west San Jorge (Coward and Cronan, 1987; Biliki, 1988).

There have been several cruises to search for placer gold off the Solomon Islands (Glasby, 1986). It was believed that gold might be transported to the sea from the Gold Ridge area of central Guadalcanal. However, although some offshore samples contained detectable amounts of gold, the results did not appear to be very encouraging from a mining point of view (Glasby, 1986).

Heavy mineral bearing sands occur on beaches and in beach–dune systems at Sigatoka, Fiji (Green, 1970), and also in the offshore area (Holmes, 1981). These are predominantly titaniferous iron sand deposits averaging 94% Fe_3O_4, 6% TiO_2 and comprising 5.5% of the total sand present.

Additional deposits of heavy mineral sands in the south-west Pacific have been reviewed by Glasby (1986). He concludes that extensive placer deposits would not be expected off many islands in the south-west Pacific. The principal reasons for this are the widespread development of coral reefs which dissipate the wave energy needed to concentrate the placers on beaches, and frequent lack or small volume of placer mineral bearing source rocks in the hinterland. This latter is especially true on coral islands and atolls, and can be held to apply to other tropical island settings around the world.

3.5 RESOURCE POTENTIAL OF EEZ PLACER MINERAL DEPOSITS

Several factors affecting the resource potential of marine placers have been touched upon in the previous pages. It is very difficult to generalize on the resource potential of these deposits because they are so variable in nature and so scattered in distribution. Local circumstances often play a large part in determining the economic viability of a deposit. Most offshore placer mining operations have been rather small scale, with the exception of the south-east Asia cassiterite mining, and even this is fragmented, illegal as well as legal mining taking place. Tin suffered a sharp setback in 1985 when the International Tin Council's support operation collapsed, and in 1990 had still not fully recovered. Nevertheless, the tin producing countries of south-east Asia are still carrying out exploration for offshore tin, in anticipation of the market's eventual recovery.

Currently one of the most active areas of offshore placer mineral development is in the field of noble metals. Increases in the price of

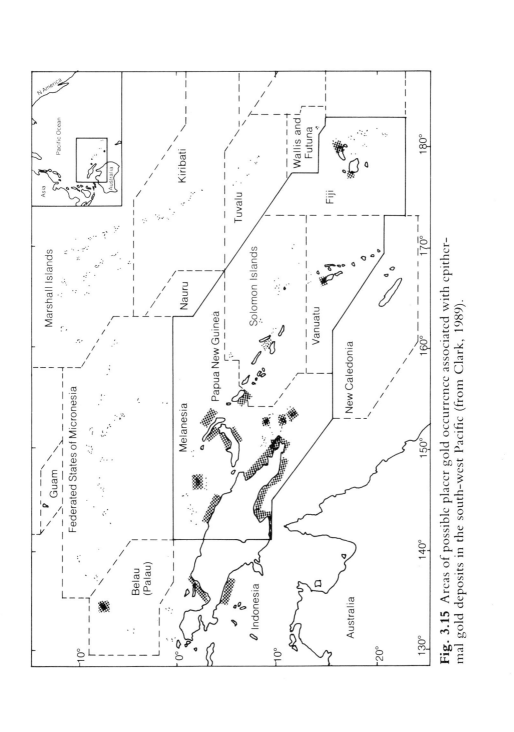

Fig. 3.15 Areas of possible placer gold occurrence associated with epithermal gold deposits in the south-west Pacific (from Clark, 1989).

gold over the past few years have led to a number of offshore deposits which were considered to be uneconomic or submarginal to be worth mining. The recent activity off Nome, Alaska, mentioned earlier is a good case in point. It is possible that developments in offshore placer mining technology based on noble metal mining may have spin-offs that are applicable to less valuable placer deposits such as cassiterite and light heavy mineral sands, which may result in these being mined more economically than hitherto.

According to Clark (1989), placer gold derived from the erosion of epithermal and porphyry gold deposits in the western Pacific and south-east Asia may have an especially great untapped resource potential. More than one thousand epithermal gold deposits are known in the region, many of which occur nearshore. Through erosion, these and porphyry occurrences will have contributed some fine grained gold to rivers and the adjacent coastal areas. Assuming that only 10% of an amount of gold equivalent to known on-land reserves in the region has been concentrated offshore, Clark (1989) estimates that there is about 200 tonnes of placer gold there. Areas where such deposits might occur are shown in Figure 3.15.

Clark (1989) also points out that some of any gold supplied to the offshore area in the western Pacific will be very fine grained to colloidal in size. Some of this could be trapped in concentrating environments such as reefs and mangrove swamps, and these could be among the most favourable sites for placer gold if they occur in association with epithermal or porphyry gold deposits onshore. Lum (personal communication, 1990) believes that the gold potential of mangrove swamps near epithermal gold occurrences may have been overlooked in the past both as a result of the very fine grain size of the gold and the difficulty in analysing for it accurately in the presence of large amounts of organic carbon.

In 1988 there were some remarkable increases in the price of certain light heavy mineral sands. Zircon, for example, of which almost all is produced from heavy mineral sands, has been the subject of increasing demand for a range of applications. According to Gooding (1988), this mineral has traditionally been regarded as a second-class placer which for most of the time has been in over-supply. However, Kenmore Resources, a Dublin based company, estimated an annual shortfall of about 100 000 tonnes of zircon, and that demand was likely to outpace supply until 1994 (Gooding, 1988). Within the near future, demand has been forecast to grow at about 5% a year.

Titanium and the minerals from which it is produced, rutile and ilmenite, have also recently been in short supply. Titanium is used mainly in the production of pigments, the consumption of which increased by 5.5% annually over the five years to 1988 (Gooding, 1988). Titanium is also used in metal alloys, especially in aircraft.

4

Precious Coral

Precious corals are not minerals in the strict sense of the term, being living coelenterates prior to their recovery, but because they can be made into jewellery, like gold and platinum, it is not out of place to consider them here.

Of the many species of coral known, relatively few are 'precious corals'. The latter are pink, black, gold and bamboo varieties, and of these pink and black corals are the most common.

Pink coral (*Corralium* sp) occurs mainly in the Mediterranean and the Pacific, and consists of high magnesium calcite (Grigg and Eade, 1981; Harper, 1988). According to Harper (1988), Mediterranean pink corals occur at shallower depths than Pacific varieties (5–300 m). In the north Pacific, they are preferentially located in two depth ranges, 2–500 m and 1000–1500 m. According to Grigg and Eade (1981), in their distribution in the Pacific they occupy habitats from Japan, along the Philippines, across to Hawaii and in the south-western Pacific, and have recently been dredged in the equatorial Line Islands region (Cronan *et al.*, 1987). They tend to grow on limestone or basalt substrates and are best developed on rises, seamounts or gently sloping current swept terraces which are free of sediment. Both high sedimentation rates and steep slopes are inhibitive to pink coral growth, the former by smothering the corals and the latter by permitting downslope transport of detritus that can abrade and topple the growths. Temperature does not appear to be critical for pink coral growth, as they grow in waters of 9°–18°C (Eade, 1980). Growth rates tend to be about 1 cm/year and they grow to sizes ranging up to 1 m with base diameters 3–10 cm. Apparently pink corals occur in commercial quantities only in small beds (Grigg and Eade, 1981).

According to Carleton and Philipson (1987), black corals (*Antipatharia* sp) occur in tropical, semi-tropical and temperate areas, but are harvested only in tropical regions such as the Caribbean, the Indo-Pacific area and off the Hawaiian Islands. They occupy much shallower waters than many pink corals, often between 20 and 80 m, and have been reported in depths of as little as 2 m off Tonga and less than 10 m off Fiji (Harper, 1988). Like pink corals, they prefer hard current-swept substrates, and often grow under overhanging rocks as their larvae avoid bright sunlight (Grigg and Eade, 1981). Common trunk diameters are in the range of 1–2 cm and heights may reach several metres (Harper, 1988). Growth rates are in the range of 5 cm/year under optimal growth conditions (Grigg, 1977). The main factors that define the quality of black coral are colour, hardness, parasitization, diameter, length and blemishes.

Gold (*Gerardia* sp) and bamboo (*Acanella* sp) corals are rarer than pink or black varieties. Gold corals are largely confined to Alaska and Hawaiian waters (Grigg and Eade, 1981), and occur in water depths of 300–400 m similar to that of the intermediate depth pink corals (Harper, 1988). Bamboo corals also occupy the 300–400 m depth range.

The precious coral jewellery industry grew up in the Mediterranean, but now both precious coral recovery and jewellery manufacture are largely based in the Pacific. The locations of the main harvesting and processing areas are shown in Table 4.1 in which it can be seen that the Philippines dominate the production and Taiwan the processing of black coral, and Japan the production and Taiwan the processing of pink coral. An important area of pink coral recovery is the Emperor Seamounts north of Midway Island. According to Glasby (1986), the total value of coral sold in 1980 was about $US 50 million. However, since then prices have escalated, and indeed pink coral increased in price by about 100% in 1986 alone (Harper, 1988). In 1989, prices of over $6000/kg for coral in an unfinished condition were common.

4.1 PRECIOUS CORALS IN THE TROPICAL PACIFIC

A number of cruises have been mounted in the Pacific by CCOP/SOPAC and other bodies in search of precious corals, with a number of recoveries being made (Harper, 1988) (Figure 4.1).

In the Cook Islands, there were 37 CCOP/SOPAC dredgings/dives between 1977 and 1988, during which there were some pink coral recoveries in relatively shallow water. However, slopes appear to be too steep to support commercial-sized beds of precious coral at the depths surveyed. Dredging in intermediate depth waters (100–700 m) in the northern Cook Islands (Manihiki, Penrhyn and Rakahanga Islands) (Figure 4.1) gave poor results, suggesting little

Table 4.1. Location of precious coral harvesting and processing areas (from Harper, 1988)
Estimated black coral harvesting and processing, 1986 (after Carleton and Philipson, 1987)

Location	Production (tonnes)	Processing (tonnes)
Philippines	60	10
Sri Lanka	20	5
Hawaii	3	1
Fiji	1	1
PNG	1	1
Tonga	5	2
Taiwan	–	70
Total	90	90

Estimated pink coral harvesting and processing, 1986 (after Carleton and Philipson, 1987)

Location	Production (tonnes)	Processing (tonnes)
Mediterranean	20	23
Japan	} 72	32
Taiwan		97
Midway Beds	60	–
Total	152	152

potential for commercial quantities of precious coral at the locations studied. Black coral finds have been reported (Harper, 1988).

Off Fiji there had been 16 CCOP/SOPAC dredgings for pink coral up to 1988, without any commercially viable recoveries being made. However, black corals have been harvested in Beqa Lagoon and surrounding waters, and worked locally. According to Harper (1988), black coral also occurs off Taveuni.

In the Solomon Islands region there have been over 200 dredgings conducted by CCOP/SOPAC for intermediate depth precious corals, during which there were 39 finds (Harper, 1988). However, most specimens were miniature compared with the Hawaiian and Emperor Seamount beds.

Black coral occurs in Tongan waters and is both worked locally, and exported. Indeed, Tonga is the principal exporter of black coral in the South Pacific region (Table 4.1). There have been numerous pink coral surveys in Tongan waters, but with little success. All the specimens recovered were small.

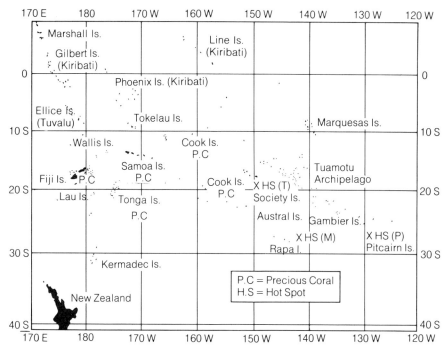

Fig. 4.1 Locations of precious coral dredgings and hotspot hydrothermal activity in the South Pacific: T = Teahitia, M = Macdonald, P = Pitcairn.

Off Western Samoa four surveys for intermediate depth pink coral suggest a reasonable potential for these deposits in the 100–600 m depth range (Harper, 1988). Pink coral was recovered at 18% of the dredgings.

It can be seen from the work done so far in the south-west Pacific that few commercially interesting precious coral beds have been located there. Since 1977, CCOP/SOPAC personnel have made over 500 dredgings or dives, mostly at intermediate depths, and the main coral type recovered has been pink coral, but in non-commercial amounts. These results suggest that the potential for finding commercial quantities of intermediate depth pink coral in the south-west Pacific are limited (Harper, 1988). Although pink coral has been recovered at many locations, environmental conditions at intermediate depths do not seem to be favourable for growth of this coral type to large size. Also the steep slopes around many of the islands are not conducive to the development of commercial-size beds. By contrast, the potential of the region for black coral appears to be good, as several beds are known and worked in Fiji, Tonga and

Papua New Guinea (Table 4.1) and others have been found throughout the region although not developed (Harper, 1988). Problems of confidentiality preclude widespread dissemination of information on shallow water precious coral finds. Although there have been no reported surveys for deep water pink coral, according to Harper (1988) the potential for its discovery appears to be good because they are known to occur in the region and there are large areas in the appropriate depth range of 1000–1500 m. Some deep water pink corals have been recovered by accident during cobalt-rich crust dredging operations.

Hawaii supports an important precious coral industry where, until recently, a submersible was employed by Maui Divers Inc in the recovery of pink and black coral from near inshore. Use of submersibles in precious coral recovery has obvious advantages over some of the more random collection methods such as dragging tangle nets over the beds, but is very expensive. This was one factor that led to the abandonment of submersible use off Hawaii, and now much of the precious coral processed in Hawaii is bought in from abroad. However, the possibility of employing an unmanned ROV in precious coral recovery was being seriously considered by Maui Divers Inc in 1989 (Slater, personal communication, 1989).

__5

Phosphorites

Phosphorite is a mixed phosphate carbonate deposit, the principal mineral in which is a variety of apatite called carbonate fluorapatite, or francolite. This mineral often occurs in the deposits in the form of pellets or nodules, giving rise to the common term phosphorite nodules.

5.1 NATURE AND OCCURRENCE

Phosphorite nodules and pellets are usually structureless but can have a layered internal structure, or be conglomeratic or oolitic in internal form. The grain size of phosphorite pellets is generally in the range of 0.1 to 0.3 mm, whereas the phosphorite nodules can get up to several centimetres in diameter. The latter are normally dark coloured brown to black, largely as a result of the presence of organic impurities. Other impurities can include clay minerals, detrital mineral grains, silica, glauconite and iron oxides. Various types of skeletal material can also be present.

Carbonate fluorapatite is a mixed calcium phosphate carbonate mineral in which fluorine can substitute. The carbonate groups and phosphate groups are mutually interchangeable, thus leading to a considerable range in the phosphorus content of the mineral (Cronan, 1980). As a result of this and the presence of impurities, the maximum P_2O_5 content of phosphorites is only about 30%, or about 14% elemental phosphorus.

The most common depth range of offshore phosphorites is less than 1000 m, often much less. They tend to occur on the middle and outer parts of continental shelves rather than in very nearshore areas, but nevertheless all occur within Exclusive Economic Zones. They are principally found off western margins of the continents, but also occur in other areas including off some eastern continental margins (Figure 5.1). They can also be locally abundant on oceanic seamounts in association with calcareous and volcanic rocks. Modern continental margin phosphorite deposition is believed to be taking

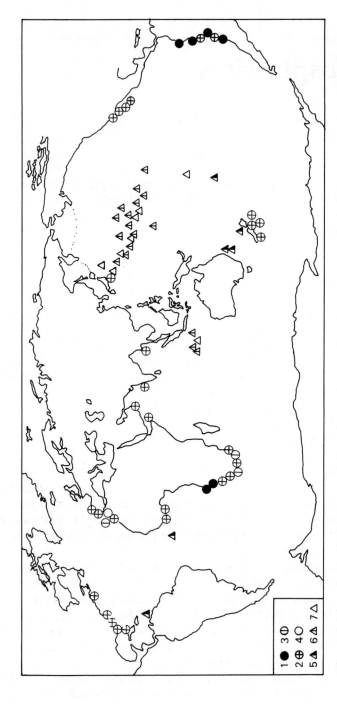

Fig. 5.1 Phosphorite locations on the sea floor: 1–4 = phosphorites on continental margins; 5–7 = phosphorites on seamounts. Ages: 1 = Holocene; 2,5 = Neogene; 3,6 = Paleogene; 4,7 = Cretaceous (from Cronan, 1980, after Bezrukov and Baturin, 1976).

place off Baja California, Peru-Chile, South West Africa, western India and eastern Australia and possibly in the Gulf of California and off North Carolina (Burnett, oral communication, 1990).

5.2 OCEANOGRAPHIC SETTINGS

One oceanographic feature which is often associated with the formation of western continental margin phosphorites is upwelling. Upwelling is the phenomenon whereby cold nutrient-rich (including phosphorus) waters are brought to the surface from intermediate depths resulting in a high level of biological productivity (mostly diatoms) in the surface waters. When these organisms die, a rapid flux of organic material to the seafloor takes place. This phenomenon has a considerable influence on the mechanism of formation of the deposits, as will be described later. However, phosphates also occur in non-upwelling situations, particularly those on some eastern continental margins, and the origin of these deposits is uncertain at the present time.

As mentioned, phosphorites also occur on seamounts in areas far from continental margins (Figure 5.1). Some of these were probably islands in the past (Cullen and Burnett, 1987) and their phosphate deposits could be analogous to those found on atolls and islands in the Pacific. The latter deposits often originate from bird droppings (guano), although other mechanisms are possible (see below), and have been mined for phosphates for more than 100 years. The low islands and atolls produced about 1 million tonnes of phosphate, mostly during the nineteenth century, but these are now largely worked out. By contrast, at the time of writing, Nauru was still producing phosphate although the deposits are largely diminished and are not thought to last much beyond the end of the present century. Although small in volume, seamount phosphorites are of interest for two reasons. First, small scale mining for local fertilizer utilization on nearby islands is possible, and second, their ages and distributions may provide information for palaeogeographic reconstruction of phosphorite bearing provinces (Sheldon, 1980).

5.3 ORIGIN OF SEAFLOOR PHOSPHORITES

The environments of modern phosphorite formation, and the origin of the deposits in them, has been reviewed at length by various authors and has been the subject of a special issue of *Marine Geology* (Burnett and Froelich, 1988). It need not therefore be considered in

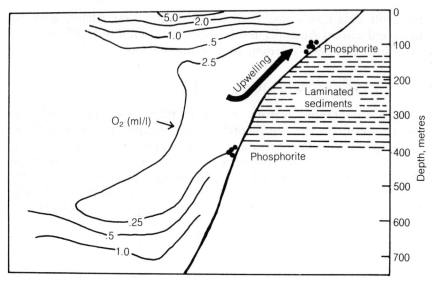

Fig. 5.2 Schematic diagram of the continental shelf and upper slope off Peru. Phosphorite tends to be concentrated at the upper and lower boundaries of the O_2 minimum zone with a zone of non-bioturbated laminated sediments in between (from W.C. Burnett).

detail here. A principal requirement is a supply of phosphorus to the seafloor, one source of which is that supplied by marine organisms, particularly diatoms, in areas of high biological productivity. However, dissolution of fish debris and/or the presence of a phosphorus-bearing microbial mat on the seafloor may also be important in this regard (Froelich *et al.*, 1988). Several theories have been advanced to account for the mode of formation of the phosphorite deposits in such situations, including the direct inorganic precipitation of phosphorus in the uppermost interstitial waters of sediments, and second, the substitution of carbonate ions by phosphate ions in calcium carbonate in the sediments to form a replacement deposit. These two mechanisms would both seem to operate through the interstitial waters of the sediments and are not mutually exclusive.

Perhaps the best studied example of a modern continental margin phosphorite deposit is that off Peru (Burnett, 1977; Burnett and Froelich, 1988). There it is proposed that precipitation of phosphorus took place in the oxygen deficient interstitial pore waters of organic rich sediments. The recent deposits appear to be largely concentrated at the margins of the seawater oxygen minimum zone where it impinges on the seafloor (Figure 5.2). This may be the result of organisms at the margins of the oxygen minimum zone preventing nodule burial by bioturbation, while the absence of organisms in the laminated oxygen deficient zone would favour burial. However, some relict phosphorite is found in the laminated zone, which was probably formed during periods of different seawater dissolved O_2 distribution in the past, and some minor pelletal phosphorite formation is also possible there (Burnett, personal communication, 1990). The conversion of such disseminated phosphorites to deposits of potential economic value is attributed to the selective concentration of the deposits, their chemical upgrading, and the removal of the barren sediment with which they are associated (Baturin, 1982). This may involve long periods of time and include emergence, erosion, redeposition and rephosphatization of the deposits. However, recent work on phosphorite formation off Peru has suggested that pelletal phosphorites can accumulate without reworking (Baker and Burnett, 1988).

According to Burnett (1987), seamount phosphorites are typically phosphatized limestones which are often coated with ferromanganese oxide crusts. In their gross composition and mineralogy, they are generally similar to continental margin phosphorites, and the amounts of phosphorus, calcium and fluorine present are found to coincide with the degree of phosphatization. However, in contrast to continental margin deposits, the seamount phosphorites are low in organic carbon and uranium, and there is an absence of pyrite in them. A phosphorite-ferromanganese oxide crust association on seamounts has been held to indicate a possible genetic link between these deposits (Halbach *et al.*, 1982; Hein *et al.*, 1985a).

Interestingly, although phosphorite deposits on Pacific seamounts appear to vary little from other phosphorite deposits in terms of their major element geochemistry, many show considerable differences from Pacific island phosphorites (Burnett, 1987). The island samples approximate closer to pure carbonate fluorapatite than do the seamount samples. This calls into question possible links between island phosphorites and seamount phosphorites which imply that the latter are merely submerged varieties of the former, although in some instances this does appear to be the case (Cullen and Burnett, 1987).

It has been known for some years that many seamounts in the central Pacific are submerged islands. Although it seems proven that some phosphate deposits on seamounts could have formed on islands prior to submergence (Cullen and Burnett, 1987), it is unlikely that the majority of them originated in this fashion (Burnett, 1987).

Guano is comprised of bird droppings, and although birds existed as early as the mid-Mesozoic, their major development did not occur until the Cenozoic era, probably in the Eocene period (Beerbower, 1968). It follows, therefore, that some seamount deposits are too old to have originated from bird droppings. Furthermore, Burnett (1987) points out that guano deposition cannot account for the origin of some phosphorites on north-western Pacific seamounts because there have been two stages of development of the phosphatization, the latter of which is believed to have occurred in a relatively deep water environment (Heezen *et al.*, 1973). Finally, Rougerie and Wauthy (1989) point out that the presence of zinc and fluorine, not present in guano, in some seamount deposits precludes an avian origin for these deposits.

In order to account for the origin of phosphorites on some atolls in the Pacific, Rougerie and Wauthy (1989) have proposed an 'endo-upwelling' process within the atoll edifices (Figure 5.3). This involves the geothermally driven upward transport of nutrient-rich deep ocean water through the fractured and porous structure of the atoll which sustains a high level of biological productivity in closed or restricted lagoons where the waters exit. In support of this hypothesis, Rougerie and Wauthy (1989) present data showing dissolved phosphate concentrations in some Polynesian lagoon waters increasing with lagoon closure. In uplifted atolls where lagoons are completely enclosed, phosphate concentrations 10–20 times those of surrounding ocean waters were recorded. It is believed that high phosphorus contents in lagoon waters could lead to even higher concentrations in the interstitial waters of the underlying sediments, resulting in the precipitation of apatite in them. According to Rougerie and Wauthy (1989) the formation of phosphorites by this process results from diagenesis of the phosphatized sediments. An additional mechanism proposed for the origin of lagoon phosphorites is direct concentration by lagoonal organic matter. Trichet and Rougerie (1990) have reported up to 1.3% P_2O_5 in lagoonal sediments on Niau (French Polynesia). Further, Burnett (1990) has reported the precipitation of carbonate fluorapatite from pore waters in sediments of a salt water lake on the island of Eil Malk in Palau. In this case, however, the rapid rate of deposition of associated sediments prevented the formation of a phosphorite deposit. Nevertheless, Burnett (1990) considers that such deposits could form either by weathering and concentration of the carbonate fluorapatite after the lake disappears, or the interaction of phosphate enriched interstitial lake waters with underlying limestones.

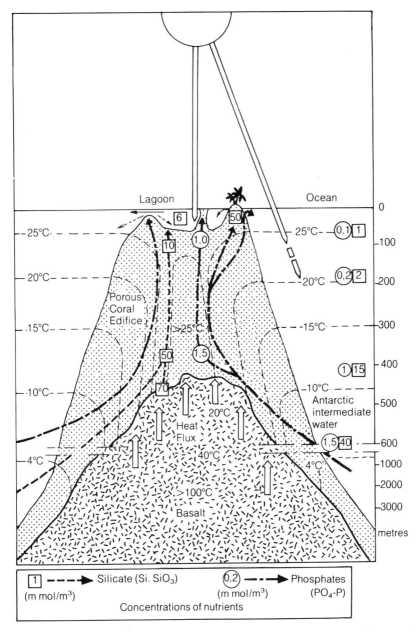

Fig. 5.3 Geothermal endo-upwelling, the geothermally driven internal ocean water circulation in an atoll (from Rougerie and Wauthy, 1989).

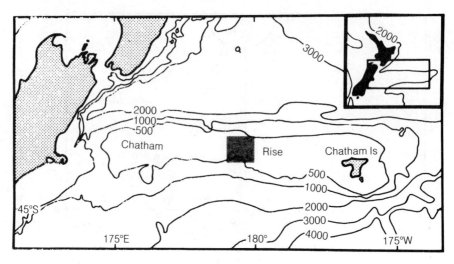

Fig. 5.4 Area of phosphorite occurrence on the Chatham Rise, off New Zealand (after Cullen *et al.*, 1981).

5.4 CASE HISTORIES

Three case studies have been selected to illustrate the variability of EEZ phosphorites. One is off New Zealand, probably to date the best studied potentially economic offshore phosphorite, the second comprises many deposits scattered among South Pacific islands, and the third is a largely buried deposit off the USA.

5.4.1 Phosphorite deposits on the Chatham Rise off New Zealand

New Zealand's economy is largely agriculturally based, and is therefore dependent upon an adequate supply of fertilizer to maintain its agricultural productivity. One of the most essential fertilizers that New Zealand has imported in large quantities is phosphorus, which is usually applied to the soil in the form of a soluble superphosphate. New Zealand, having no on land phosphorite deposits of its own, finds the import of phosphate fertilizer in large quantities a necessity which utilizes valuable foreign exchange.

The phosphorite deposits on the Chatham Rise off New Zealand (Figure 5.4) have been suggested as being an alternative source of phosphate which could reduce New Zealand's reliance on imported fertilizer. In order to evaluate this possibility, a joint New Zealand/

West German research programme was carried out in 1981 to investigate the possibility of the Chatham Rise deposits being mined at some time in the future.

According to Cullen (1986), the Chatham Rise phosphorite is a loose nodular gravel, sometimes more than 70 cm thick, which is intermixed on the seafloor with glauconitic sandy muds and muddy sands. It rests on Oligocene chalk. Individual fragments of phosphorite vary in size from 15 cm across down to a few millimetres, and are composed of indurated and phosphatized pelagic limestone. Phosphatization is thought to have occurred in mid- to late Miocene times between 7 and 9 million years ago. Phosphorites average about 9.4% elemental phosphorus which is slightly higher than the phosphorus content of superphosphate fertilizers. Indeed, Cullen reports experiments demonstrating that Chatham Rise phosphorite performs at least as well as superphosphate on direct application to the soil without prior processing.

In their distribution, the phosphorite deposits occur along the crest of the Chatham Rise between Reserve Bank, and the western shelf of the Chatham Islands (Figure 5.4). Throughout this entire area, the concentration of phosphorite shows a patchy distribution. In the most promising areas, the phosphorite concentration varies from about 10 to 15% of total weight, and the sediment above the calcareous ooze or chalk contains approximately 80 kg of phosphorite per square metre (Cullen, 1986).

According to von Rad and Kudrass (1984), the genesis of the Chatham Rise phosphorite has involved a multi-staged process (Figure 5.5). It started with the formation of a chalk hard ground surface in middle- to late Miocene times, which became bored and fragmented during a long period of non-deposition and erosion. This formed chalk pebbles which became slightly ferruginized. One interesting feature of the Chatham Rise phosphorite reported by von Rad and Kudrass is the close spatial relationship of phosphorite rich areas containing an average concentration of $66 \, \text{kg/m}^2$ phosphorite to phosphorite poor areas with an average coverage of only $11 \, \text{kg/m}^2$. This phenomenon was explained in terms of a morphologically controlled multi-stage phosphatization of the ferruginized chalk pebbles during several phases of submarine erosion and hard ground formation in late Miocene times (Figure 5.5). The uplifted parts of fault blocks favoured phosphatization because they were better exposed to phosphatizing processes involving overflow by phosphorus-rich waters, than deeper lying areas which were not so exposed and suffered a higher rate of pelagic sedimentation. Each erosion-phosphatization event enhanced the phosphorite content of

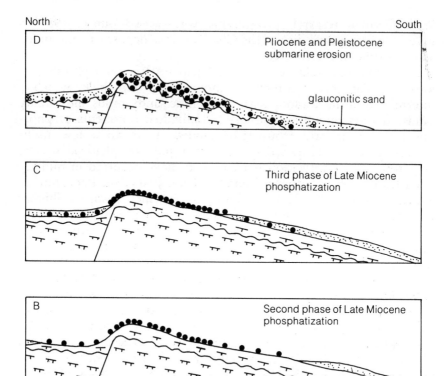

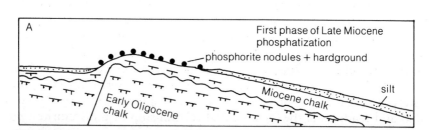

Fig. 5.5 A model for the morphological control of phosphatization on Chatham Rise with repeated cycles of phosphatization, erosion and hardground formation (from von Rad and Kudrass, 1984).

the elevated areas, thereby increasing the difference between those areas and the adjacent ones.

5.4.2 Seamount phosphorites in the south–west Pacific

Cullen (1986) has provided an overview of phosphorite occurrences in the south–west Pacific. Deposits of phosphatized coral occur on

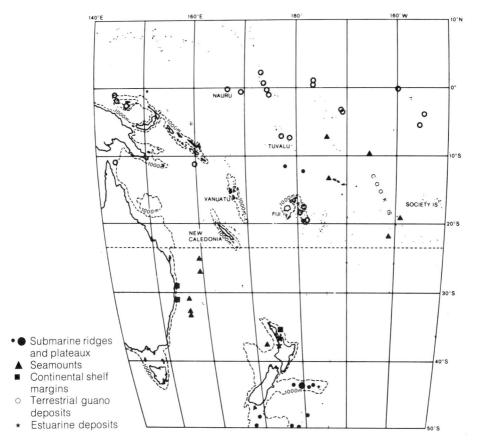

Fig. 5.6 Distribution of phosphatic deposits in the south-west Pacific region (from Cullen, 1986).

the tops of various seamounts, guyots and submarine ridges there (Figure 5.6). Its incidence appears to become more frequent as the equator is approached.

Cullen and Burnett (1986) have recovered phosphatized foraminiferal limestone and coral from depths between 1000 and 2000 m in the region of the Tokelau and Northern Cook Islands (Figure 5.6). Large scale subsidence of volcanic piles seems to be the explanation for the presence of coralline material at such depths. In the vicinities of Western Samoa and Rarotonga, phosphate occurs in ferromanganese concretions. Similar material has been recovered near Aitutaki. The presence of submerged phosphorites on Solomoni and McLeod guyots which resemble insular phosphorite on Nauru and Ocean Islands are believed to be sunken island phosphorites (Cullen and Burnett, 1987).

According to Cullen (1986), phosphatic mineralization has been reported on some of the 'Tasmantid' guyots in the western Tasman Sea, and on Gifford Guyot and Capel Bank that rise from the northern part of Lord Howe Rise. The phosphorite occurs in two forms – as phosphatized limestone, and as phosphatic veneers on volcanic cobbles. Phosphatic material, composed of fluorapatite, hydroxyapatite and calcite has also been reported on the crest of Aotea Seamount near New Zealand. Cullen (1986) suggests this phosphate represents the modified surface rind of a limestone that forms a capping or partial capping to the seamount.

Cullen (1986) has noted that the nature of the parent carbonate material of the phosphorite varies throughout the south-west Pacific. He considers this to be partly related to water temperature and hence to latitude. In low latitudes, the phosphorite frequently consists of altered coral or coral-reef debris, whereas in cooler latitudes the parent carbonate consists of pelagic foraminiferal ooze or molluscan, echinoderm, bryozoan or algal skeletal fragments.

The relationship between terrestrial guano deposits on islands and submarine phosphates on south-west Pacific seamounts in the south-west Pacific has already been mentioned. In view of the subsidence of some seamounts, it is possible that some of the phosphates on south-west Pacific seamounts could have formed through sinking of guano deposits. In addition, the dissolution of guano deposits can lead to local enrichment of phosphates in oceanic waters, and this could have promoted local phosphatic replacement of calcareous accumulations on seamounts (Cullen, 1986). Elsewhere in the south-west Pacific, however, there appears to be no question of the major involvement of guano in phosphate deposition on seamounts. In these instances, Cullen (1986) envisions a process that combined minimal deposition or erosion with localized upwelling of phosphate-rich cold bottom waters, Rougerie and Wauthy (1989) have proposed the 'endo-upwelling' concept described earlier, and Burnett (1990) and Trichet and Rougerie (1990) invoke the role of lagoonal organic matter.

5.4.3 Phosphorite deposits on the south-eastern United States continental margin

A model for the occurrence and origin of largely buried phosphorite off the south-east United States has been developed by Riggs (1987) and other works by Riggs referenced therein. The phosphorite formation began about 29 million years ago and continued until about 25 million years ago. After a pause it began again about 19 million years ago and continued in a cyclical manner until about 13 million years ago. In addition, small concentrations of phosphorite formed between 5 and 4 million years ago during the Pliocene period.

The Miocene period was one during which phosphorite bearing sediments formed in abundance throughout the world. Those off the south-east United States (Figure 5.1) formed during this episode. Major Miocene deposits such as those in North Carolina and Florida occur on the emerged coastal plain and are well known geological-ly. However, these onland deposits are thought to be just the edge of much larger formations which extend seawards under the continental shelf (Riggs, 1987). Occasionally, seafloor outcrops of Miocene sediments occur where the grains of phosphorite can be reworked into younger sediments. In some places, the phosphorus content of recent sediments closely follows the phosphorite content of the underlying Miocene strata (Riggs, 1987). In addition, Pevear and Pilkey (1966) and Gorsline (1963) have shown significantly increased phosphorus concentrations in recent sediments on the inner continental shelf off the south-east United States where it has been demonstrated by Woolsey (1976) that phosphorite rich Miocene strata occur below the sea floor. The potential for finding new phosphorite deposits within Miocene strata off the south-east United States is therefore quite good (Riggs, 1987).

According to Riggs (1984), the formation of the phosphorites off the south-east United States can be related to multiple depositional sequences formed within Miocene and Pliocene sediments in response to transgressive cycles of sea level change. Variations in Gulf Stream processes through each transgression superimposed complex and changing patterns of deposition and erosion on the cyclical sedimentation. During periods of transgression, phosphorite bearing sediments were formed, while during regression and low stand portions of sea level cycles, non-depositional scouring and channelling took place together with diagenetic alteration of the deposits previously formed. As a result of these processes an abnormal amount of phosphorus was deposited in the Miocene sediments off the south-east United States, along with contem-poraneous deposition of glauconite, diatomaceous muds, organic matter, dolomites and magnesium rich clays (Riggs, 1987).

5.5 RESOURCE POTENTIAL OF EEZ PHOSPHORITES

World food demands are rising as population increases, and fertilizer supplies will have to rise in order to keep pace with the increased production of basic food crops. The growth of fertilizer production and consumption since the Second World War has been unmatched at any time in the past. Both production and export levels in the mid-1970s were five times as high as in the mid-1950s (Burnett and Lee, 1980).

While the resource potential of phosphorites lies principally in their use as a fertilizer, they have other uses such as in the chemical industry for the manufacture of elemental phosphorus and phosphoric acid, and also for the minor elements that they contain such as vanadium, uranium, fluorine and the rare earth elements (REE), which may be obtained from them as by-products.

All of the increased phosphorite production in the past decades has been met from land deposits of phosphorite and much of future increases will also be met from this source. However, in certain situations offshore phosphorite production may be necessary to ensure a continuing supply of fertilizer. Use of offshore phosphorite may be brought about by at least three factors:

1. depletion of some land phosphorite supplies (e.g. Ocean and Christmas Islands);
2. great distance of some markets from land sources of phosphorite, and offshore phosphorites locally available, e.g. some oceanic islands, New Zealand (see below); and
3. increasing pressures on phosphorite bearing lands for non-mining purposes (e.g. USA).

Marine phosphorites can thus be divided into three classes in terms of their resource potential:

1. Deposits adjacent to well-developed countries which may be mined in the future in competition with land deposits, as are sand and gravel in some places at the present time. Deposits off the south-east United States might fall into this category.
2. Deposits occurring off countries which might find it more economic to mine locally offshore than to import from overseas. Deposits off India, New Zealand and some Pacific Islands might fall into this category.
3. Deposits occurring off countries with large deposits on land and which have little use for phosphorites themselves. These would probably not be exploited within the foreseeable future. The deposits off Morocco would fall into this category.

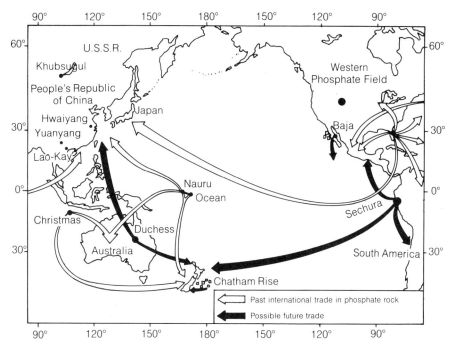

Fig. 5.7 Phosphate occurrences and supply routes in the Pacific (after Burnett and Lee, 1980).

The Pacific region provides a good example of some of the problems of phosphorite mining and supply. Within the Pacific region, phosphorite is the most critical fertilizer raw material, and thus the resources and flows of phosphates are of great importance there. Phosphorite deposits within the Pacific region are shown in Figure 5.7. In the 1970s, all the countries of the Asia-Pacific region together produced only about 8% of the world's phosphate supplies, while the region contains 20% of the world's accessible land area and contained 52% of its population (Burnett and Lee, 1980). The importance of phosphate rock is not just restricted to the less developed countries of the Pacific. Japan imports phosphate from Florida, while Australia and New Zealand recently consumed almost the entire outputs of Nauru and Christmas Island (Indian Ocean), the latter now exhausted. Such long supply routes (Figure 5.7) result in high transportation costs, and could, at least, help to lead to the Chatham Rise deposits off New Zealand being brought into production. Exploration for phosphorite bearing lagoons will also be stepped up, as is already happening in parts of the South Pacific.

It is within this framework of perceived future difficulties in the supply of phosphates within the Pacific region that the Chatham Rise deposits off New Zealand have been given serious consideration as a phosphorite source. According to Cullen (1979), New Zealand yearly imported about 1 million metric tonnes of phosphate rock from Nauru and Christmas Islands in the 1970s. Since Nauru, like Christmas Island, will probably be depleted before the end of this century, the Chatham Rise phosphorites lying within New Zealand's EEZ are considered as a potential alternative source of phosphate fertilizer. Kudrass and Cullen (1982) have reported the average P_2O_5 concentration in Chatham Rise phosphorites to be 22%, which, however, is low compared with terrestrial or island deposits. This low grade phosphorite is not considered suitable for production of super-phosphate fertilizer but it dissolves in the acid environment of many New Zealand soils and can be used as a direct application slow release fertilizer (Mackay *et al.*, 1980). Further, the average K_2O content of 1.5%, present in the glauconitic rim of the deposits, could partly contribute to their fertilizing effect. Cullen (1979) reports that further advantages of the Chatham Rise phosphorite are savings by low costs for crushing and granulation, and by the reduced frequency of application compared with superphosphate.

More recently, however, Falconer (1989) has reappraised downwards the resource potential of Chatham Rise phosphorite, concomitant with a decline in fertilizer imports to New Zealand since 1985 due to the depressed state of its agricultural economy. The company that held the exploration licence has allowed it to lapse, although there has been interest from other companies. Market forces have led to a general reduction of interest in the Chatham Rise deposits, partly because of the scale of the operation that would be needed to mine them and because direct application phosphorite can be imported relatively cheaply into New Zealand. Nevertheless, Falconer (1989) points out that should phosphorite prices rise sufficiently or there be supply difficulties, the Chatham Rise deposits could become economic, especially if they were mined only periodically with the mining ship being employed elsewhere at other times.

Manganese nodules

Manganese nodules are not primarily Exclusive Economic Zone deposits, having their major abundance in the deep oceans (Figure 6.1). However, important deposits of manganese nodules have been discovered in Exclusive Economic Zones in recent years and these may well be mined when International Seabed Area nodule mining starts, or possibly even before in view of their not being subject to the Law of the Sea Convention and their having a close proximity to land areas which can exert jurisdiction and ownership over them. One of the main justifications for including them in this work is that the Exclusive Economic Zone nodule deposits have recently attracted considerable attention, both by the United States and some of the South Pacific island nations within whose Exclusive Economic Zones the deposits occur. This latter occurrence has prompted a considerable amount of exploration in recent years, and this is continuing at the time of writing.

Manganese nodules are concretionary deposits of iron and manganese oxides that contain significant amounts of a variety of what are normally minor and trace elements in other sea floor deposits, and have been reviewed at length by Cronan (1980). Nickel, copper and cobalt are the metals of greatest economic interest in the deposits, and reach concentrations of up to about 1.5% but not all in the same nodules. Nickel and copper covary with each other, and both vary negatively with cobalt. This means that nickel and copper-rich nodules are commonly found but these nodules are low in cobalt and conversely cobalt-rich nodules can occur, but these are generally low in nickel and copper. The average composition of nodules from different oceans is shown in Table 6.1. In general, highest concentrations of potentially economic metals occur in Pacific nodules with intermediate concentrations in Indian Ocean nodules and lowest concentrations in those from the Atlantic. A number of manganese nodule mine sites have been claimed in the International Seabed Area under the pioneer investor provisions of the Law of the Sea Convention.

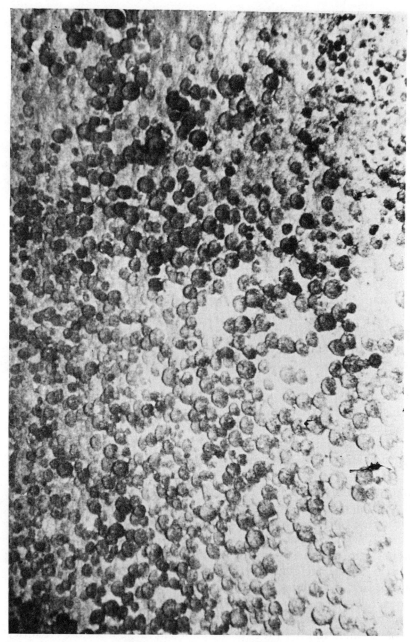

Fig. 6.1 Manganese nodules on the deep sea floor.

Table 6.1. Average partial composition of manganese nodules from different oceans (in weight %) (from Cronan, 1980)

	Pacific	*Atlantic*	*Indian*
Mn	19.78	15.78	15.10
Fe	11.96	20.78	14.74
Ni	0.634	0.328	0.464
Co	0.335	0.318	0.230
Cu	0.392	0.116	0.294
Pb	0.084	0.127	0.093
Zn	0.068	0.084	0.069

6.1 MORPHOLOGY AND STRUCTURE

The ferromanganese oxides of which nodules are composed normally accrete around a nucleus, which can be a piece of volcanic rock, organic remains, a fragment of a pre-existing nodule, or some other solid object on the sea floor. In their morphology, manganese nodules can be very variable, ranging from spherical through oblate to flattened varieties. A morphological classification of nodules is given in Table 6.2. One factor which affects the morphology of manganese nodules, and their composition, is the source from which the metals in the nodules are derived. Sources of metals to the marine environment in general have been considered in Chapter 1. In the case of the nodules, there are two main sources of metals, the continents and submarine volcanic activity. Continentally supplied metals can either precipitate directly or be concentrated by organisms and released when they decay on the sea floor. Volcanically supplied elements can either be released from submarine volcanic rock by chemical alteration processes, or be supplied by submarine hydrothermal activity. Superimposed upon these primary sources of metals to nodules are diagenetic processes which can secondarily concentrate manganese and other metals in nodules at the sediment surface by their upwards remobilization from buried sediments. Nodules which receive the bulk of their metals from seawater, irrespective of how the metals got into seawater in the first place , are termed hydrogenetic or hydrogenous nodules and nodules which receive a substantial proportion of their metals as a result of diagenetic cycling through the interstitial waters of the sediments underlying them are referred to as diagenetic nodules. Most nodules contain a mixture of hydrogenetic and diagenetic phases, as will be discussed in more detail when the origin of nodules is considered.

Table 6.2. Morphological classification of manganese nodules (from Meylan, 1974)

Prefix: s = small   = <3 cm ⎫
       m = medium = 3–6 cm ⎬ nodule size (maximum diameter)
       l = large   = >6 cm ⎭
Primary morphology: [S] = Spheroidal
                      [E] = Ellipsoidal
                      [D] = Discoidal (or tabular–discoidal form)
                      [P] = 'Poly' (coalespheroidal or botryoidal form)
                      [B] = Biological (shape determined by tooth,
                                  vertebra, or bone nucleus)
                      [T] = Tabular
                      [F] = Faceted (polygonal form due to angular
                                  nucleus or fracturing)
Suffix: s = smooth (smooth or microgranular) ⎫
       r = rough (granular or microbotryoidal) ⎬ surface texture
       b = botryoidal ⎭
Examples: l[D]b   = large discoidal nodule with botryoidal surface
          m–l[E]$^{s}_{r-b}$ = medium to large ellipsoidal nodules with smooth
                              tops, rough to botryoidal bottoms

---

Many nodules are non-uniform in their morphology. The so-called 'hamburger' nodules (Figure 6.2) from the north-eastern equatorial Pacific are an example of this phenomenon. These have flattened tops and rounded undersides, reflecting a variable supply of metals from both above and below. This illustrates an important point about the supply of metals to manganese nodules in that not only can these be variable on an ocean-wide basis, but can also vary on the scale of a single nodule (Figure 6.2). Thus, in the case of the 'hamburger' nodules, the upper surfaces are receiving metals from seawater and therefore consist of hydrogenous phases, whereas the lower surfaces are receiving metals from the interstitial waters of the sediments and therefore are composed of diagenetic phases.

In addition to variable hydrogenetic or diagenetic supply of metals to the nodules, other factors can also affect their morphology. These include the shape of the nucleus in that it determines the initial shape of the nodule on commencement of growth and can often influence its shape throughout its growth history. Another factor which can affect the morphology of manganese nodules is fracturing of pre-existing nodules on the sea floor. This sets a limit to the size to which nodules can grow, and indeed the fractured portions often serve as nucleii for new nodule growth, thus increasing the density of nodule

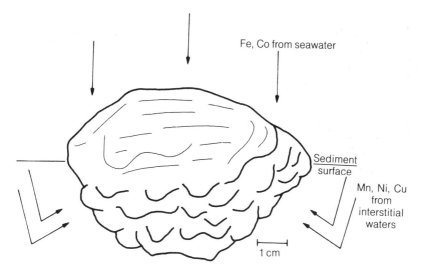

Fe, Co from seawater

Sediment surface

Mn, Ni, Cu from interstitial waters

1 cm

Fig. 6.2 Morphological and compositional differences between the top and bottom of a 'hamburger'-shaped nodule from the north-eastern equatorial Pacific (from Cronan, 1978).

coverage in any one area. The actual cause of the fracturing is not always entirely clear, but is thought to result from either shrinkage of the nodules consequent on post-depositional reorganization of their interiors, or as a result of their movement by organisms, or by other factors to be outlined below.

The main internal structural feature of manganese nodules, other than the nucleus, is the concentric banding which characterizes them. The banding results from the accumulation of layers of different reflectivity. The more highly reflective layers are generally richer in manganese than the more poorly reflective ones, which tend to be richer in iron (Figure 6.3). This layering superficially resembles growth rings in a tree, and has been held by some authorities to be directly analogous to such rings in that each layer represents a period of growth under conditions distinctively different from the adjacent layers. However, this may not be entirely the case as there is evidence for internal reorganization of manganese nodule interiors (Cronan and Tooms, 1968; Roy *et al.*, 1990) which could considerably influence the nature of the growth rings that are seen in the nodules. Further, growth rings are not always continuous features around a nodule, particularly in those nodules in which the upper portion is hydrogenous and the lower portion is diagenetic and which therefore have received metals from differing sources (Figure

Fig. 6.3 Concentric internal banding in a manganese nodule (from CNEXO).

6.2). In these cases, growth rings can pinch out at the position of the sediment water interface and/or overlap with other growth rings, thereby creating small unconformities within the growth structure of the nodules (Sorem, 1967).

On a smaller scale, a major structural feature of nodules is the non-ferromanganese oxide framework in which the oxides reside (Murray and Renard 1891; Lalou reported in Cronan, 1980). This framework supports the nodule structure and is composed largely of silicate and aluminosilicate minerals. Within this non-ferromanganese oxide framework, the ferromanganese oxide 'segregations' often show concentric banding themselves down to the highest levels of magnification resolvable by microscopic techniques. It is difficult to account for these structures in terms of primary growth, and together with the non-ferromanganese oxide framework, they may result from diagenetic reorganization of nodule interiors after primary deposition of colloidal or amorphous phases (Cronan and Tooms, 1968; Burns and Burns, 1978, Cronan 1980; Roy *et al.*, 1990). However, it is clear that this is not always the case, nor indeed may it be commonly so, as many of the internal features of nodules can be ascribed to primary precipitation mechanisms, particularly where there is evidence of organic associations (Greenslade, 1975); biological remains are a common constituent of both nodule surfaces and their interiors.

6.2 GROWTH RATES

The rates of growth of manganese nodules can be determined either by dating their nucleii, or by measuring age differences between their different layers. These techniques have indicated that most deep sea nodules grow slowly in the order of a few millimetres per million years (Table 6.3). However, some nodules appear to grow much faster than the rates given in Table 6.3. Evidence for this is to be found in features such as unconformities suggesting discontinuous growth, and biological features suggesting variable growth rates on different surfaces of the nodules. In general, the hydrogenetic nodules tend to grow more slowly than do nodules to which there is a significant diagenetic supply of metals. This can best be exemplified by comparing the upper and lower surfaces of hamburger-shaped nodules shown in Figure 6.2. The upper hydrogenetic portion of the nodule has grown much more slowly than the lower diagenetic portion, and this is exemplified by the larger size of the lower portion of the nodule than the upper.

Evidence that some nodules grow very fast indeed is provided by their having accreted around human artefacts such as spark plugs, iron nails, fragments of naval shells, and beer cans (Goldberg and Arrhenius, 1958; Moorby, 1978; Cronan, 1980). However, these nodules are of little economic value, being very poor in nickel, copper and cobalt.

Two main factors appear to influence the rate of growth of nodules, the rate of supply of their constituents and catalytic processes on the precipitation surface. In situations where there is a rapid local supply of nodule forming metals, such as submarine volcanic activity, or an important diagenetic source, rapid precipitation of ferromanganese oxides can occur. By contrast, over much of the deep ocean floor, far from local sources of elements or localized concentration processes, normal slow hydrogenous precipitation processes prevail. The rate of catalytic oxidation of ferromanganese oxides on the nodule surface may be influenced by the nature of the surface. Metallic surfaces, for example, tend to

Table 6.3. Growth rates of selected manganese nodules (after Ku *et al.*, 1977)

Sample	Latitude	Longitude	Depth (m)	Growth rate (mm/10^6 yr)	Method
North Atlantic:					
C58–100	30°57'N	65°47'W	4800	4	^{230}Th, ^{231}Pa
A226–41	30°59'N	78°15'W	830	<2	^{230}Th, ^{231}Pa
G74–2374	30°31'N	79°01'W	876	<2	^{230}Th, ^{231}Pa
G74–2384	30°53.5'N	78°44'W	843	<2	^{230}Th, ^{231}Pa
Lusiad AD4	6°03'N	32°22'W	1020	8–10	^{230}Th, K-Ar
South Atlantic:					
V16–T3	13°04'S	24°41'W	4415	3	^{230}Th, ^{231}Pa
North Pacific:					
FanBd–20	40°15'N	128°27'W	4500	3	K-Ar
V21–D2	35°54'N	160°19'W	5400	4	^{230}Th, ^{231}Pa ^{234}U
V21–71a	27°54'N	162°31'E	5870	2.5	^{230}Th
V21–D4b	14°25'N	145°52'W	4618	3	^{230}Th, ^{231}Pa
6A	19°39'N	113°44'W	4000	4	^{230}Th, ^{231}Pa
Carr 5	9°26.5'N	113°16.5'W	3700	17–24	^{230}Th, ^{234}U
MP 26	19°N	171°W	1464	10	^{230}Th
Zetes–3D	40°16'N	171°20'E	3300	0.8–2.3	^{10}Be, ^{230}Th
Tripod–2D	20°45'N	112°47'W	3000	4	^{10}Be, ^{230}Th
Dodo–9D	18°6'N	161°50'W	5500	2.1	^{230}Th
Dodo–15–1	19°23'N	162°20'W	4160	1.8	^{10}Be
Wah24F–8	8°18'N	153°03'W	5143	10 (3 nods.)	^{230}Th
2P–52	9°57'N	137°47'W	4930	7.3	^{230}Th
South Pacific:					
V18–D32	14°18'S	149°32'W	2000	3.1	^{230}Th
V18–T119a	12°27'S	159°25'W	5000	6	^{230}Th, ^{231}Pa
V18–T119b	12°27'S	159°25'W	5000	3	^{230}Th, ^{231}Pa
DWHD–47	41°51'S	102°01'W	4240	1–6	K-Ar, ^{230}Th
DW72	21°31'S	85°14'W	920	18	^{230}Th
2P–50	13°53'S	150°35'W	3695	1	^{230}Th
TF–1, TF–2	13°52'S	150°35'W	3623	1 (2 nods.)	^{230}Th
E24–15	35°48'S	134°50'W	4696	6.4	^{230}Th
DO23G	near Tuamoto I.		1600	3.5	^{230}Th
Pacific:					
J1	–	–	–	10.7–15.1	^{230}Th
H1d	–	–	–	6.8	^{230}Th
G1d	–	–	–	3.6–10.3	^{230}Th

Table 6.3. Cont.

Sample	Latitude	Longitude	Depth (m)	Growth rate (mm/10^6 yr)	Method
Indian Ocean:					
V16–T19a	29°52'S	62°36'E	4500	2.8	^{230}Th
V16–T19b	29°52'S	62°36'E	4500	2.9	^{230}Th
V16–T19c	29°52'S	62°36'E	4500	2.3	^{230}Th
Antarctic:					
E17–36	55°S	95°W	4700	3	^{230}Th, ^{231}Pa
E5–4	60°02'S	67°15'W	3475	4–19	^{230}Th

accrete thicker coatings of ferromanganese oxides faster than non-metallic surfaces.

6.3 DISTRIBUTION

The distribution of manganese nodules is very variable both on a local and on an ocean-wide scale. First, it is proposed to review the major ocean-wide features of nodule distribution and the factors affecting it, and then local variations and the factors affecting them will be considered separately.

In the Pacific Ocean, nodules are quite abundant, generally more than 10 kg/m^2, in an area between about 6°N and 20°N extending from about 120°W to 160°W, the so-called Clarion-Clipperton Zone (Figure 6.4). Most of this area is not within any Exclusive Economic Zone but small parts of it occur within the Exclusive Economic Zones of Clarion Island and Clipperton Island which are Mexican and French respectively, and these areas have received some study by those nations (e.g. Carranza-Edwards *et al.*, 1987; Rosales Hoz and Carranza-Edwards, 1988). The Clarion Island EEZ abuts one of the mine sites of the International Consortia, and thus should contain potentially economic nodules. The limits of the area of high nodule abundance in the Clarion-Clipperton Zone are largely determined by sedimentation rates. To the north and east, detrital sediments from the continents inhibit nodule growth, while to the south, rapid carbonate sedimentation associated with the equatorial zone of high biological productivity has a similar effect.

In the South Pacific, nodule distribution is more irregular than in the North Pacific, possibly because of the greater topographic

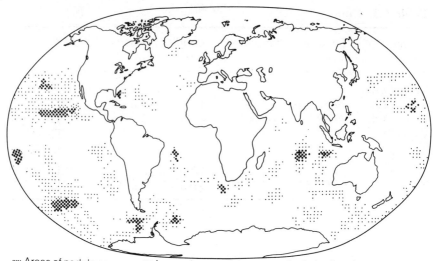

::::: Areas of nodule coverage ✖ Areas where nodules are particularly abundant

Fig. 6.4 Distribution of manganese nodules in the World Ocean (modified from Cronan, 1980).

diversity of the South Pacific and the abundance of island groups scattered throughout it. The main areas of high nodule abundance in the South Pacific are, like in the north Pacific, areas where sedimentation rates are low, away from the major land masses. However, locally high nodule abundances occur in the outer parts of some of the Exclusive Economic Zones of the central south Pacific island territories (see case histories). The small size of the islands, and thus the limited amount of detrital material they are able to supply to their adjacent sea floor, results in sedimentation rates in the outer parts of the Exclusive Economic Zones being little different from such rates in the central south Pacific at large. These nodules have been the subject of a considerable amount of exploration activity in recent years (Cronan *et al.*, 1991). The distribution of nodules in the South Pacific is shown in Figure 6.4.

In the Indian Ocean, nodules are most abundant to the south of the equator, in the basins to the east and west of the 90°E ridge. The areas that show moderate to high nodule coverage include the Central Indian Basin, the Wharton Basin, the South Australian Basin, the Madagascar Basin, the Crozet Basin, the Agulhas Plateau and the Mozambique Ridge and Channel (Figure 6.5). Little of any of these areas occurs within Exclusive Economic Zones, and thus the nodules in them fall under international jurisdiction. However, the

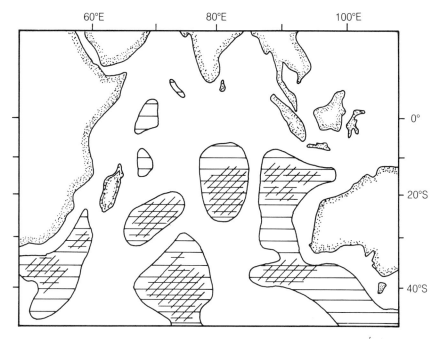

Fig. 6.5 Areas of moderate to high nodule abundance (crosshatched) in the Indian Ocean.

western margin of the Central Indian Basin is bounded by the Chagos Laccadive Ridge, which includes the Chagos Islands. The Exclusive Economic Zone of the Chagos Islands extends into the Central Indian Basin manganese nodule belt, adjacent to one part of the Indian mine site, and therefore should contain some deposits of potential economic value. Like in the Pacific Ocean, sedimentation rates are low in the areas of high nodule abundance in the Indian Ocean.

In the Atlantic Ocean there appears to be lower concentrations of nodules than in the other two oceans (Cronan, 1975), probably because of its relatively high sedimentation rates. Greatest abundances occur in the deep water basins on either side of the Mid-Atlantic Ridge where sedimentation is inhibited, and under the circumpolar current.

As indicated, the main factor affecting nodule distribution is sedimentation rates. These are largely controlled by two processes, sediment supply rates and bottom currents. Where sedimentation is rapid, nodules are rare. This is because the sediments bury the nodule nucleii or embryonic nodules before they can grow to an appreciable

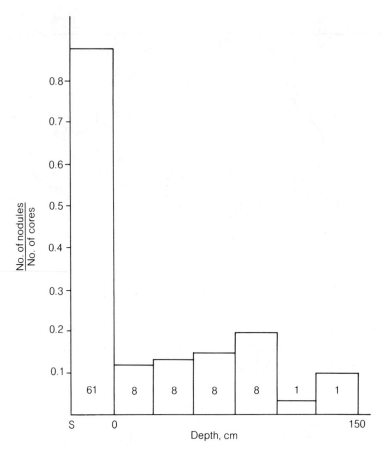

Fig. 6.6 Distribution of nodules with depth in Pacific Ocean gravity cores. The numbers at the foot of each column refer to the actual number of nodules found in each depth interval. The column marked 'S' represents surface nodules only (from Cronan and Tooms, 1967a).

size. Where sedimentation rates are low, either because of a limited supply of sediment or because vigorous bottom currents prevent deposition, nodules can be abundant. Although sedimentation rates are the main factor affecting nodule distribution and abundance, other factors that can affect these parameters are the presence or absence of suitable nucleii on the seafloor to initiate nodule growth, their age, and their proximity to sources of metals.

An apparent paradox in regard to nodule distribution is why the nodules are concentrated at the sediment surface when in many cases the average rate of deposition of their associated sediment is greater than that of the nodules themselves. In some cases this may be due to sediment erosion and redistribution, but in many cases it appears that burrowing organisms maintain the nodules at the sediment surface either by nudging them or by burrowing underneath them and thereby moving sediment under them (Sanderson, 1985).

While occurring in greatest concentrations at the sediment surface, nodules also occur buried within sediments (Figure 6.6). Various estimates of their abundance within buried sediments have been made (Cronan, 1980) and most conclude that they are about half as abundant within the upper metre of sediments as in the surface layer. However, the distribution of buried nodules within sediments is not always random, sometimes layers of buried nodules are found which could represent ancient erosion surfaces. It has been shown by von Stackelberg and Beiersdorf (1987) that buried nodules in the Clarion-Clipperton Zone often represent embryonic varieties of the more mature nodules found at the sediment surface, and are compositionally similar to the interiors of those nodules. Evidently, the surface nodules represent concretions whose growth and upward movement has kept pace with sedimentation rates, whereas the buried nodules represent embryonic nodules which got left behind for one reason or another during the general upward movement of nodules with continuing sedimentation.

6.4 LOCAL VARIABILITY IN NODULE DISTRIBUTION

From just a cursory glance at the literature on manganese nodules, one could be forgiven for getting the impression that large areas of the deep ocean floor are uniformly covered with nodules. This is far from the truth. Even in areas which have high average densities of nodule coverage, their local abundance can vary from zero to more than 50% coverage of the sea floor over short distances (0–30 kg/ m²). This local variability in nodule abundance will have important implications when it comes to actually mining manganese nodules, both within and outside of EEZs.

One of the earliest reports of local variability in nodule abundance (Cronan, 1967) was on manganese nodules from part of the Indian Ocean. Quite large differences in the abundance, morphology and composition of nodules from sites only a few kilometres apart were reported. Mineral phases in the nodules also differed. The topography of the area where the nodules were found was rather rugged and it was concluded that topographic variations, by influencing the environment of deposition, were responsible for the variations in abundance and composition noted. Similarly, Bezrukov and Skornyakova (1976), drew attention to the local variability of nodules in areas of rugged topography. They found nodules distributed irregularly in relation to a volcanic topography of over 2000 m relief in a small area of the south-western Pacific.

Although a local variability in nodule abundance had been established in areas of strong topographic contrast by the early 1970s, little was known about the possibility of local variability occurring in areas of more uniform seafloor topography, such as that prevailing in the Clarion-Clipperton Zone in the north-eastern equatorial Pacific. Up until the mid-1970s, this area had been sampled at a low station density except by mining companies whose data were not released. With increased commercial and academic interest in the area, once its ore potential had been realized, much more detailed sampling took place and hitherto unexpected variations in nodule abundance were found in it.

Much of the variability found in the abundance of ore grade nodules in the Clarion-Clipperton Zone appears to be related to small scale variations in topography, the deposits occurring pre-ferentially on certain parts of hills or in certain parts of valleys, dependent on local circumstances. On one abyssal hill only 250 m high, Andrews and Friedrich (1980) recorded considerable variability in the abundance of nodules. Close to the top of the hill a sparse nodule population occurred, and volcanic rock outcrops were common. Volcanic rock outcrops were also found on the slopes of the hill but between them dense populations of nodules occurred. Nodules were fairly sparse near the bottom of the hill, and sometimes buried nodules were found to occur. Andrews and Friedrich (1980) also recorded local variability in nodule abundance in other areas, in one of which the nodule distribution followed topographic trends. In the centre of a valley and on the crest of a hill there occurred the lowest nodule abundances, whereas the highest abundances were found on slopes and near the sides of the valley.

Other workers have also drawn attention to variations in nodule abundance with topography, sometimes producing conflicting re-ports. For example, it was noted by Moore and Heath (1966) that in part of the Central Pacific nodules are usually less abundant in valleys and at the bases of slopes than on hilltops and on slopes. By contrast, it was found in the DOMES area of the Clarion-Clipperton Zone that nodules were most abundant on slopes and at the bases of hills. Also in the DOMES area, it has been found that the nodules are most abundant where the uppermost acoustically transparent layer is less than 15 m thick, indicating a low rate of sedimentation (Piper *et al.*, 1978). Similar findings by Mizuno and Moritani (1976) demonstrated an inverse correlation between nodule abundance and the thickness of the transparent layer in the Central Pacific to the west of the Clarion–Clipperton Zone.

Small-scale sampling in the Central Pacific Basin with free-fall photograbs, piston cores, box cores, coupled with seismic investigations have shown a characteristic pattern of local variation of nodule facies around abyssal hills (Usui *et al.*, 1987). Most of the nodule variability previously described regionally in the central Pacific is also found on the scale of kilometres in this hill area. In general, nodules are abundant on and around the hills where the mean sedimentation rate has been very slow or non-existent from late Tertiary to the Quaternary, but nodules are scarce in basin areas off the hills where the sedimentation rates are rapid.

It is evident from these reports that nodule distribution in regard to small scale topographic features can be quite variable. Several factors affect it and there are few generalizations that can be drawn in this regard. It appears that the effect of topographic variation on the localized distribution of nodules, when it occurs, is most likely to be through its influence on bottom currents, sedimentation rates, sediment redistribution and erosion.

6.5 MANGANESE NODULE MINERALOGY

The mineralogy of manganese nodules is exceedingly complex, and its terminology is fraught with confusion. The nodules contain a number of different phases, usually very fine grained and often intergrown. In addition amorphous material can be abundant.

There appear to be two main manganese phases in manganese nodules, one exhibiting a 9.7 Å reflection, and the other a 2.4 Å reflection. In addition, some nodules exhibit a phase with a 7.1 Å reflection, but this is thought by some workers to be an artefact of drying the samples prior to X-ray diffraction analysis. The terminology of the phases exhibiting these reflections has undergone several modifications since they were first described in the 1950s (Buser and Grutter 1956). In many respects, the original terminology of Buser and Grutter is still the best in that it makes no comparisons with terrestrial manganese minerals but simply records the X-ray diffraction data. Hence the terms 10 Å manganite, 7 Å manganite and δMnO_2 for phases showing the 9.7, 7.1 and 2.44 Å reflections respectively still has much to commend it. Other terminologies based on comparisons with terrestrial minerals have been the subject of much as yet unresolved controversy, and therefore for the purpose of the present work the original terminology of Buser and Grutter is

Table 6.4. Typical XRD lines for manganese nodules

δMnO_2	10Å manganite	7Å manganite
	9.68	7.27
	4.80	3.60
2.44	2.46	2.44
1.42	1.42	1.41

Table 6.5. Compositional variability of manganese nodules dominated by different mineral phases

Phase	10Å manganite	δMnO_2
Enrichments	Mn, Ni, Cu, Zn	Co, Pb

preferred. The X-ray diffraction data for these phases, the principal basis of their identification, are given in Table 6.4.

Nodules dominated respectively by the 10 Å manganite and 2.4 Å manganite phases show considerable compositional differences. Those containing 10 Å manganite as their main phase are enriched in manganese, nickel, copper and zinc, and indeed some workers consider that these elements in their divalent states are needed to stabilize the 10 Å manganite structure. Those nodules containing δMnO_2 as their main phase are enriched in iron and cobalt (Table 6.5). These differences probably largely result from the variable substitution of the metals in the different structures of the minerals concerned, but adsorption phenomena may also influence variable trace metal uptake by mineralogically different phases (Calvert and Price, 1977). The compositional differences between the 10 Å manganite and δMnO_2 rich nodules are extremely important in terms of understanding nodule geochemistry as they demonstrate a strong mineralogical control on the chemical composition of the nodules.

Nodule mineralogy varies regionally throughout the oceans and with different environments of deposition. In the Pacific, for example, 10 Å manganite occurs predominantly in the east of the Basin whereas δMnO_2 is more common in the more topographically dissected western areas and is particularly abundant in shallow water deposits. The causes of this differentiation have not been clearly established, but are thought to include the degree of oxidation of the environment of deposition, and the nature of the sediments being

Table 6.6. Approximate variability in the concentrations of economically important elements in manganese nodules (in weight %)

	Max.	*Min.*
Mn	30.28	5.41
Fe	26.32	4.36
Ni	1.68	0.136
Co	2.57	0.045
Cu	1.64	0.028

deposited in association with the nodules and diagenetic reactions in them.

It is evident that as a result of chemical differences between nodules of differing mineralogy and regional variations in nodule mineralogy throughout the oceans, regional mineralogical variations in nodules will be one factor affecting regional chemical variations in them. This will be discussed in more detail below.

6.6 COMPOSITIONAL VARIABILITY

Nodules are enriched in a whole variety of elements over the normal abundances of those elements in the Earth's crust. Four of the most economically valuable metals – manganese, nickel, copper and cobalt – vary very much indeed (Table 6.6), and it is only when they are present in concentrations at the high end of their concentration range that the nodules are of potential economic value.

Some elements show considerable compositional variations with depth, nickel and copper tending to increase in concentration with increasing depth while cobalt tends to decrease. There are at least two possible reasons for these variations. First, they could be related to mineralogical variations in nodules with depth, as it has previously been pointed out that nodules of differing mineralogy are of differing composition. Second, they could be related to biogenic dissolution phenomena at and below the Carbonate Compensation Depth (CCD) liberating elements into the bottom and interstitial waters of the sediments where they are available for uptake by forming nodules. Such biogenic dissolution phenomena would involve mainly the soft parts of sunken organic remains and faecal material as it has been shown that nodule forming elements are not particularly concentrated in skeletal material (Chester, 1990), and would be of greatest importance under areas of high biological

Table 6.7. Approximate partial composition of manganese nodules from different environments (in weight %)

	Abyssal	Seamounts	Active ridges	Continental margins
Mn	16.78	14.62	15.51	38.69
Fe	17.27	15.81	19.15	1.34
Ni	0.540	0.351	0.306	0.121
Cu	0.370	0.058	0.081	0.082
Co	0.256	1.15	0.400	0.011

productivity. For example, elements such as nickel and copper which are concentrated in biological material and which are liberated at and below the CCD, appear to be preferentially taken up in 10 Å manganite bearing nodules forming at that depth and, indeed, may, as mentioned, be stabilizing the structure of that mineral. The dissolution of the organic material may also be lowering the degree of oxidation of the bottom environment such that the 10 Å manganite phase can form.

As might be expected from the mineralogical and depth related variations in nodule composition described above, nodules from different environments on the seafloor show considerable compositional differences. Cronan (1977) defined seven environments in which distinctively different compositional varieties of manganese nodules occurred. However, in general the greatest compositional differences are found between deposits from four of these environments, continental borderlands, seamounts, active ridges and the abyssal seafloor (Table 6.7). Continental borderland deposits are rich in manganese of diagenetic origin and low in most other elements. Many of these are Exclusive Economic Zone deposits but because of their low minor element content they are of little or no economic value as far as being a source of metals is concerned. However, the high manganese content of these deposits could make them suitable for use as a catalyst in emission control procedures (Siapno, personal communication, 1989). Seamount deposits are rich in cobalt, possibly due largely to their being rich in the δMnO_2 phase which has the ability to concentrate cobalt into its structure. In many respects, these nodules are similar to cobalt-rich crusts which are the subject of the next chapter. Active ridge nodules have high iron/manganese ratios, possibly due to volcanic/hydrothermal sources of iron and a lack of diagenetic sources of manganese owing to the generally thin or non-existent sediment cover on which the deposits rest. Abyssal nodules are highest in nickel and copper, possibly due to their containing 10 Å manganite as an important mineral phase and which has a structure suitable to accommodate these metals, and also because they probably receive these elements from the dissolution of sinking organic remains.

Even within the abyssal environment, there are considerable regional variations in the composition of manganese nodules. For example, nickel and copper tend to be highest in the deposits in the north-eastern equatorial Pacific and parts of the south-eastern and south-central Pacific and to decrease generally both westwards and to the north and to the south. A major control on this regional variability seems to be biological productivity in the overlying waters in that it influences both the supply of metals to the seafloor available for uptake by the nodules and, by organic decay on the seafloor, influences the diagenetic reactions by which the metals are actually taken up within the nodules. Under the equatorial zone of high biological productivity, there is an abundant flux of metals and organic carbon to the seafloor which participates in nodule forming reactions. The organic carbon is most concentrated in sediments at or near the CCD where calcium carbonate is no longer able to dilute it, and it is near this depth that the highest nickel and copper concentrations in the nodules occur. Both above and below the CCD, its abundance decreases either due to its dilution by $CaCO_3$ or its decay in the water column (Cronan, 1989a). The biological supply of metals to the seafloor, together with organic carbon, the decay of which helps to enrich the metals in the nodules, is one of the two factors of greatest importance in leading to nickel and copper rich nodules in the north-eastern equatorial Pacific. The second factor is the presence within these nodules of the 10 Å manganite phase which, as mentioned, has a structure suitable to accommodate these metals. Such a combination of circumstances, however, is not unique to the north-eastern equatorial Pacific but also occurs in the central Indian Ocean and in parts of the south equatorial Pacific (see case histories below) where similar environmental conditions prevail. What this indicates is that the chemical composition of manganese nodules is extremely sensitive to the depositional environment, and this has proved useful in developing exploration models for the nodules both within and outside of Exclusive Economic Zones (Cronan, 1984).

6.7 LOCAL COMPOSITIONAL VARIABILITY

Just as there are local variations in nodule abundance super-imposed upon large scale variations in abundance over wide areas of the oceans, local compositional variations in the deposits are superimposed upon regional variations in their composition. Many of these variations appear to be related to small scale variations in seafloor relief. Bezrukov and Skornyakova (1976) noted strong compositional variations in an area of volcanic topography of over 2000 m relief in the south-west Pacific. This was an area in which they also noted large variations in nodule abundance, as mentioned above. Friedrich and Pluger (1974) noted compositional variations in nodules in a small area of the north-east equatorial Pacific. They found high manganese/iron ratios and enrichments of nickel and copper in deposits from the north of the area whereas deposits from the south have lower manganese/iron ratios and lower concentrations of copper and nickel. Similar variations in nodule composition had been reported by Calvert *et al.* (1978). Halbach and Ozkara (1979) have drawn attention to the observation that local variations in nodule composition can be correlated with local variations in mineralogy. This was also noted by Cronan and Tooms (1967b) for nodules within a small area on the flanks of the Carlsberg Ridge in the Indian Ocean. Halbach and Ozkara (1979) noted in part of the Clarion-Clipperton Zone that nodules containing the mineral δMnO_2 rest on the upper slopes of the abyssal hills and have lower manganese/iron ratios and higher cobalt contents than do nodules containing 10 Å manganite which rest on the lower slopes of hills and which are enriched in manganese, nickel and copper. Andrews and Friedrich (1980) observed similar features in an area that they studied in that iron decreased in amount with increasing distance from topographic highs, while manganese, nickel and copper increased.

Causes of small scale topographically related local variations in nodule composition in the Clarion-Clipperton Zone area seem to be related, at least in part, to the path through which elements reach the deposits. In particular, whether the elements are precipitated from seawater or from the interstitial waters of the sediments underlying the nodules seems to be of particular importance in this regard. In the former situation, the deposits appear to contain δMnO_2 as their main mineral phase and are enriched in iron and cobalt. In the latter situation they contain 10 Å manganite as their main mineral and are enriched in manganese, nickel and copper. Gradations between these two extremes can be found in nodules over quite small distances on the seafloor and as mentioned earlier, even within individual nodules. Variations in sediment accumulation rate can affect the path through which the elements reach the nodules and thus influence nodule compositional variability as well as nodule distribution.

6.8 THE ORIGIN OF MANGANESE NODULES

It is evident from the information presented in the preceding sections that no single theory of origin can account for all the variability seen in manganese nodules. Historically, there have been two main schools of thought regarding manganese nodule origin, one suggesting that they are volcanic in origin and the other that they are hydrogenous in origin (Murray and Renard, 1891). For many years, these two theories seemed incompatible, but as more information on manganese nodules has accumulated, it has become evident that both theories of origin have certain merits. It is quite clear that manganese nodules precipitate from seawater or from the interstitial waters of marine sediments, but how the metals originally got into those waters is not always so clear. Both volcanic, including hydro-thermal, and continental sources of metals to the oceans have been demonstrated (see Chapter 1) and no doubt a fraction of metals from both of these sources eventually finds its way into hydrogenous nodules. In the case of diagenetic nodules, the path of the metals is a little clearer in that they have been precipitated by diagenetic reactions at and below the sediment–water interface. Thus we have a spectrum of compositional types of nodules ranging from purely hydrogenetic, perhaps best exemplified by the ferromanganese oxide encrustations and some nodules on exposed rock substrates, to nodules which have been heavily influenced by diagenetic processes and which have received a major fraction of their metals from within the sediments. Interestingly, even nodules which have all the attri-butes of hydrogenetic phases, i.e. δMnO_2 mineralogy, low nickel and copper content, low manganese content, moderate to high iron and cobalt contents, still receive some diagenetic component. According to Aplin and Cronan (1985b), it seems unlikely that any nodules resting on a *sediment* substrate will be entirely free of a diagenetic component.

6.9 CASE HISTORIES

Two contrasting examples of manganese nodule deposits in EEZs are discussed, in order to illustrate the variability of these deposits.

6.9.1 Blake Plateau

Large deposits of manganese nodules and other ferromanganese oxide deposits such as pavements and slabs occur on the Blake Plateau (Figure 6.4), a major morphological feature occurring off

Florida within the US Exclusive Economic Zone (Riggs and Manheim, 1988). The deposits occur in water depths of 600 to 1000 m between 30.5° and 32.2°N.

The ferromanganese oxides occur in a semi-continuous belt. Between periods of phosphatization, the Gulf Stream swept the plateau producing high energy bottom currents with winnowing of fine sediment and erosion producing areas of lag gravels of phosphorite, phosphatized limestone, and silicified carbonate. These surfaces became penetrated and encrusted by ferromanganese oxides to form pavements. Smaller nuclei formed slabs and round nodules (Riggs and Manheim, 1988).

The Blake Plateau ferromanganese deposits demonstrate that nodules and crusts can form at shallow ocean depths, from normal ocean water, and in close proximity to a major land mass.

The Blake Plateau was initially estimated to contain between 10 and 100 million tonnes of nodules (Manheim *et al.*, 1982). A US Geological Survey cruise (Manheim, 1982) subsequently suggested that nodule tonnages are nearer 10 million tonnes while ferromanganese pavements may exceed 200 million tonnes distributed over some 14 000 km². The Blake Plateau nodules compositionally resemble World Ocean averages for manganese nodules in terms of manganese, cobalt and nickel concentrations; but are distinctly lower in copper, nickel and zinc than prime Pacific nodules (Haynes *et al.*, 1982). In terms of 1988 commodity prices, molybdenum, vanadium and cerium have higher potential values in the Blake Plateau nodules than copper does (Riggs and Manheim, 1988). The catalytic properties of Blake Plateau nodules have been noted by Kaufman (1976) who found that their catalytic hydrogenation capacity (in terms of conversion of benzene to cyclohexane) was high. Use of the nodules as catalytic absorbers of vanadium and nickel in petroleum refining would enhance the spent catalyst in these metals (Cruikshank, 1983, quoted in Riggs and Manheim, 1988). Riggs and Manheim (1988) conclude that the Blake Plateau represents a large resource of manganese nodules, phosphate and associated ferromanganese crusts within the US EEZ.

6.9.2 South equatorial Pacific EEZs

As mentioned earlier, some of the Exclusive Economic Zones of the south equatorial Pacific island nations have received considerable attention in recent years as potential sources of manganese nodules. Nickel and copper rich manganese nodules are located on the southern margin of the equatorial zone of high biological productivity, an area which falls partly in the region bounded by the 50 g and 100 g of carbon per square metre per year productivity isolines (Figure 6.4), near and below the CCD on sediments composed largely of siliceous ooze. Two principal environmental parameters have been taken into account in explaining the occurrence of these nodules (Cronan, 1984). First, productivity, in that nickel and copper rich deposits are most likely to occur in areas where productivity is more than $50\,\text{gC}\,\text{m}^{-2}\,\text{y}^{-1}$ (fifty grams of carbon per square metre per year) which will ensure that enough organic material is transported to the seafloor to drive the diagenetic reactions leading to the formation of nickel and copper rich nodule phases. Second, depth, and in particular a depth near or below the CCD, because it is only in that depth range that sedimentation rates are likely to be sufficiently low to enable the metal rich organic phases to become concentrated enough to promote diagenetic reactions leading to the formation of nickel and copper rich nodules. Above the CCD, as mentioned earlier, the organic material is diluted by calcium carbonate inhibiting its role as a diagenetic supplier of nickel and copper to nodules, and below the CCD more decay of organic material will take place in the water column thereby reducing the amount sedimented and able to mediate in diagenetic reactions leading to nickel and copper uptake in nodules on the seafloor.

Exploration for potentially economic manganese nodules has been carried out within the EEZs of the Cook Islands (the Aitutaki–Jarvis Transect), Kiribati and Tuvalu (Figure 6.7). The results of these surveys have been synthesized by Cronan *et al.* (1989), Cronan and Hodkinson (1989), Cronan and Hodkinson (1990), and Cronan *et al.* (1991) and are summarized below.

The abundance of manganese nodules in the Cook and Central Line Islands area is highly variable (Figure 6.8). Greatest abundances occur in the middle of the area, around 15°S, to the east of the Manihiki Plateau. In this region abundances of over $30\,\text{kg/m}^2$ are not uncommon. These abundances are substantially greater than the $10\,\text{kg/m}^2$ generally considered to be the minimum for potential nodule mining. By contrast, abundances in the equatorial part of the region between the equator and about 10°S tend to be low, often less than $5\,\text{kg/m}^2$ over large areas.

In terms of their Ni + Cu + Co content, the nodules from the Cook and Central Line Islands EEZs exhibit considerable variations. Both the copper and nickel content of the nodules increases towards the equator with grades of above 2% combined Ni + Cu mainly concentrated in a zone between 2°S and 6°S. Zinc also tends to

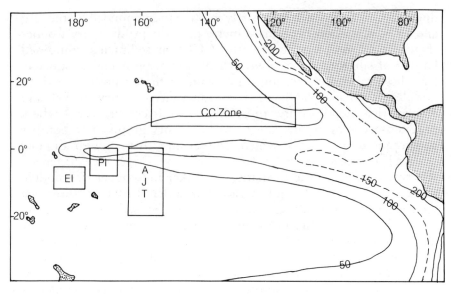

Fig. 6.7 Productivity isolines and detailed study areas for manganese nodules in the equatorial Pacific Ocean: CC Zone = Clarion Clipperton Zone; AJT = Aitutaki-Jarvis Transect; PI = Phoenix Islands, EI = Ellice Islands).

increase towards the equator, as does manganese. Conversely iron and cobalt increase from minimum values at the equator to maximum values in the Aitutaki Passage region at about 18°S.

Examination of the compositional variation in the deposits with depth, shows maximum values of nickel and copper near the CCD. Further, the depth of maximum Ni + Cu contents increases towards the north as does the depth of the CCD, supporting the view that maximum nickel and copper concentrations in the nodules are related to the zone of maximum decay of biogenic material within the sediments at and near the CCD.

Moderate to high cobalt values are found in nodules from the central part of the Cook Islands area between 10° and 15°S (Figure 6.9). This is a major new finding (Cronan, 1989a). Large areas of manganese nodules in high abundance and containing cobalt up to around 0.5% considerably extend our knowledge of the distribution of potentially economic nodule deposits. These nodules lie on a substrate of red clay or zeolitic clay, with little or no biogenic material. Hitherto, cobalt enrichment in nodules (and encrustations) has been thought to be largely confined to seamounts or other elevated areas. Although the cobalt contents of the central Cook

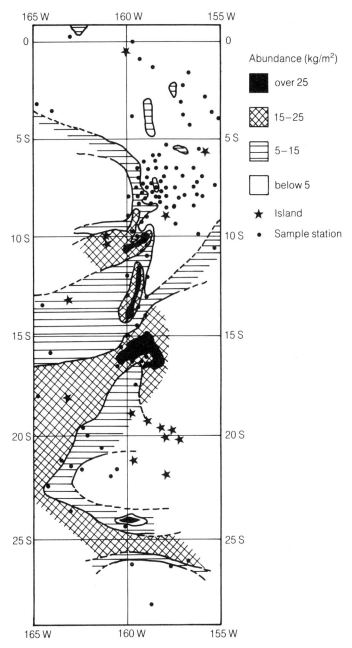

Fig. 6.8 Abundance of manganese nodules in the Cook and Line Islands area (Aitutaki-Jarvis Transect) (from Cronan *et al.*, 1989).

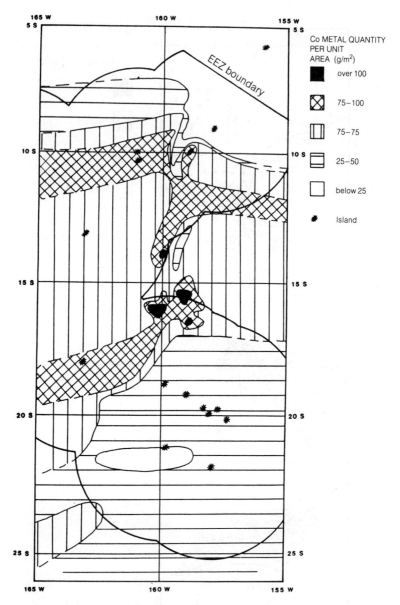

Fig. 6.9 Cobalt metal quantity in manganese nodules from the Aitutaki-Jarvis Transect (from Cronan, 1989a).

Islands nodules are not as high as in many of these deposits, they are significantly higher than the average cobalt content of manganese nodules which, combined with their large abundance, results in a very large reservoir of cobalt existing on the seafloor in this area. The resource potential of these deposits will be considered in the next section, and a comparison of them with cobalt-rich crusts will be made after the crusts themselves are described in the next chapter.

The Exclusive Economic Zones of the Phoenix Islands (Kiribati) and Ellice Islands (Tuvalu) do not occupy such a great latitudinal range as the Cook and Line Islands EEZs (Figure 6.7), and therefore do not exhibit the same range of compositional variability in the deposits as found in those areas. Nevertheless, significant variations do occur which confirm the trends observed in the Cook and Line Islands region. Highest nodule abundances in the Phoenix Islands EEZ, over 25 kg/m^2, occur in a small area in the south around 5°S, 170°W. There is a general trend towards increasing abundance with increasing latitude south of the equator. Compositionally, manganese shows a slight tendency to increase in a northerly direction in the Phoenix Islands EEZ whereas the reverse is the case for iron. Nickel and copper both tend to increase from south to north, with maximum nickel and copper values of about 1.5% each being centred more or less on the equator.

Cobalt shows the opposite trend, with maximum values of about 0.4% just south of 4°S. Manganese is at maximum in the nodules at depths around 5000–5500 m, near the CCD, as is copper and nickel (Cronan and Hodkinson, 1989).

Within the Tuvalu EEZ, manganese nodule abundances are generally low. Maximum abundance is around 10 kg/m^2 and abundances of over 5 kg/m^2 are confined to the northern part of the area only (Cronan and Hodkinson, 1990). Compositionally neither manganese, iron nor nickel shows any relationship to latitude. However, copper shows a weak increase with increasing southerly latitudes and cobalt shows the reverse. This behaviour of copper and cobalt is the opposite of what is found in the island groups to the east and illustrates a lack of influence of biological productivity on the metal values in this region compared with what happens in those island groups. Examination of the productivity isolines for the Central Pacific (Figure 6.7) shows that Tuvalu falls to the south-west of the high productivity area and therefore its manganese nodules should not be subject to productivity influences. The present data would bear out this supposition, and, incidentally, highlight the importance of biological productivity in influencing the compositional variability of nodules.

Within the depth range sampled in the Tuvalu EEZ, manganese and nickel vary little with depth, while iron and cobalt show a slight decrease with depth and copper increases with depth. In these respects, the nodules are acting like hydrogenous ferromanganese oxide crusts rather than diagenetically influenced nodules, further illustrating the lack of biological productivity influences on them.

6.10 RESOURCE POTENTIAL OF EEZ NODULES

There is a very large literature on the resource potential of manganese nodules (see Cronan, 1980, for a review of pre-1980 material) but virtually all of this relates to manganese nodules in the proposed International Seabed Area. Some of the provisions relating to the mining of ISA nodules would not relate to the mining of those in EEZs, and thus the resource potential of the deposits, if mined according to UNCLOS III provisions in the ISA, could be different in the two areas. However, these differences are not an inherent feature of the deposits themselves but of the perceived regimes under which they might be extracted, and as the latter could change substantially before manganese nodule recovery starts in earnest, many of the provisions relating to the resource potential of nodules in general will probably apply to those within Exclusive Economic Zones.

Various studies have been undertaken in order to evaluate the probable economic returns of manganese nodule mining ventures. However, the results from different studies vary widely and there is no consensus of opinion as to the degree of likely profitability of manganese nodule mining. One of the most critical assumptions in any such study is the prices of the metals that would be extracted. Over the past two decades these have varied considerably and recently have increased after a long period during which they were depressed. As mentioned above, manganese nodules contain only few metals which are considered to be of any economic interest. These are nickel, copper, cobalt, manganese and possibly zinc and molybdenum. Most manganese nodule mining scenarios envisage either a three- or four-metal operation, i.e. nickel, copper, cobalt and possibly manganese. However, it is widely accepted that the economics of manganese nodule mining should be compared with that of terrestrial nickel mining, principally nickeliferous laterites, and viewed primarily as a nickel recovery operation. Nevertheless, in recent years, the relative importance of manganese as a potentially recoverable metal has increased, particularly by consortia that contain a steel company.

According to Moncrieff and Smale-Adams (1974), any deep-sea mining operation depends upon eight factors. These are the size of the ore body, the rate of mining, the grade of the ore, the metallurgical recovery, the selling price of the product, capital costs, operating costs and legal, political and physical parameters. Most of these factors are not affected by whether or not the deposits are inside or outside of Exclusive Economic Zones and therefore one can attempt some comparisons between identified future ore reserves in the International Seabed Area and possible future manganese nodule mining sites in EEZs.

As far as the size of the ore body is concerned, the consensus appears to be that a manganese nodule mine site should be able to sustain a 1.5 to 3 million tonnes per year operation for at least 20 years. In other words, the total size of the deposit needs to be at least 30 million tonnes and preferably nearer 60 or more million tonnes. Most workers have concluded that the average abundance of nodules required for a first generation mine site is $10 \, kg/m^2$, or more with a minimum of $5 \, kg/m^2$ and that the average and cut off contents of Ni + Cu should be 2.27% and 1.18% respectively (Archer, 1979; Bastien Thiery *et al.*, 1977; Kildow *et al.*, 1976). A number of metallurgical recovery processes have been suggested for extracting metals from the manganese nodules, of which hydrometallurgical and pyrometallurgical processes appear to be the most favourable. As mentioned earlier, the selling price of the metals has varied considerably over the past few years and 1970s projections for 1980s metal prices have proved to have been over-optimistic. Thus it is very difficult to reach any consensus as to what the prices of nickel, copper and cobalt are likely to be by the end of this century, and it is possible that manganese nodule mining will not start in earnest until threshold metal values have been exceeded for some time rather than starting in anticipation of the exceeding of those values. However, it is interesting to note that the proposed French nodule mining operation, as envisaged, is projected as being profitable on the basis of rather conservative pre-1980 metal prices giving a 12–14% internal rate of return (Herrouin *et al.*, 1989). This is a more optimistic analysis than some carried out earlier by other consortia. Capital and operating costs in nodule exploitation will vary dependent upon the nature of the mining and processing techniques used, together with transport, exploration systems, research and development and other such factors. Processing is the most costly part of any nodule exploitation scenario, and thus the area where the greatest savings could be made. Use of an existing processing plant could considerably reduce costs compared to building a new one and shipboard or undersea processing could also provide cost breakthroughs. The optimistic French analysis of nodule mining mentioned above was largely the result of savings made in the processing side of the operation.

It is really only in the legal and political factors that EEZ nodules differ from those in the International Seabed Area in that the ownership of them resides with the coastal state rather than the International Seabed Authority. It is widely regarded that the deep sea mining regime proposed for the International Seabed Authority is restrictive as far as future manganese nodule miners are concerned and this may well act to the detriment of nodule mining in the deep sea area and in favour of that in Exclusive Economic Zones. The Exclusive Economic Zones of south equatorial Pacific nations contain large areas where first generation manganese nodule mine site criteria may be satisfied.

One way of attempting to assess the resource potential of manganese nodules in Exclusive Economic Zones is by calculating the amounts of metals per unit area of the seafloor there, and comparing these with values found in areas of the Clarion-Clipperton Zone which have been selected for future manganese nodule mine sites. While Ni + Cu + Co is a commonly used statistic in the evaluation of nodule grades in resource estimations, it does not take account of the independently varying prices for the metals concerned. In order partly to get over this problem, the so-called nickel equivalent value of the other metals can be calculated based on a formula which utilizes the average price ratio of nickel:copper:cobalt. Of course, this will change over any period of time, and thus so will the nickel equivalent value. Within the Exclusive Economic Zones of the Cook Islands and central Line Islands, for example, the nickel equivalent concentrations of metals based on 1988 prices were multiplied by the nodule abundance on the seafloor by Cronan *et al.* (1989) to give an estimate of nickel equivalent metal quantity per unit area. Maximum quantities were found to occur at around 15°S with generally high values between 10°–20°S. North of 10°S and south of 20°S metal quantities are fairly uniformly lower. To reach some estimate of the total amount of potentially extractable metals in manganese nodules within the Cook and central Line Islands EEZs, nickel equivalent metal quantities in all deposits present in abundances of more than $5 \, kg/m^2$ were totalled. For the area under consideration (Figure 6.8) these were 0.3 million tonnes for that part of the Line Islands EEZ considered and 109.7 million tonnes for that part of the Cook Islands EEZ considered. These were estimates of in-place quantities only and as such did not indicate either potential recoverability or mineable amounts. The much greater amounts of nickel equivalent metals in the Cook Islands EEZ than in that part of the Line Islands EEZ considered result mainly from the greater abundance and high cobalt content of nodules in the former area.

Cobalt enrichment together with high nodule abundances represents a new concept in assessing the resource potential of manganese nodule deposits (Cronan *et al.*, 1991), and has features in common with the evaluation of the resource potential of cobalt-rich crusts to be considered in the next chapter. There is no doubt that the high cobalt in the Cook Islands EEZ considerably enhances the resource potential of the nodules there. Nevertheless, assuming at least a three metal mining operation, it is the total amount of the metals present in the deposits that are of importance and which should be compared with the metals in nodule deposits in the Clarion-Clipperton Zone.

The 109.7 million tonnes of nickel equivalent nickel, cobalt and copper, in the Cook Islands EEZ compares quite favourably with amounts of metals available in nodules within potential mine sites in the Clarion-Clipperton Zone area. Assuming an average minesite area of $75\,000\,km^2$ in the Clarion-Clipperton Zone and a nodule abundance of $10\,kg/m^2$ with an average $Ni + Cu + Co$ content of 2%, one reaches a figure of 15 million tonnes combined nickel, cobalt and copper for such a minesite. On this basis, therefore, the Cook Islands EEZ could host several minesites. Although the figures are not directly comparable, they do point to a considerable resource potential for manganese nodules within the Cook Islands Exclusive Economic Zone.

Turning to the total metal quantities present in the Exclusive Economic Zones of the Phoenix Islands and Tuvalu, the resource potential of nodules in these areas appears to be lower than in the Cook Islands EEZ. Within the Phoenix Islands EEZ, Cronan and Hodkinson (1989) have reported 11.8 million tonnes of nickel equivalent metals, less than the amount present in one $75\,000\,km^2$ Clarion-Clipperton Zone minesite. In the Tuvalu Exclusive Economic Zone a similar calculation by Cronan and Hodkinson (1990) gave only 1.4 million tonnes of nickel equivalent metals present. The differences between the resource potential of nodules in the three Exclusive Economic Zones considered here relate largely to variations in the size of the areas considered, differences in the abundance of the deposits between the three areas and to a lesser extent to compositional differences between them.

Cobalt-rich nodules occur on some seamounts in the western and southern Pacific. As mentioned on p. 100, these are similar in composition to cobalt-rich manganese crusts with which they are usually associated, and indeed often represent broken crust fragments or rocks which have formed a talus on which further ferro-manganese oxide growth has taken place. For this reason, their nucleii are often disproportionately large compared with the nucleii of manganese nodules in the Clarion-Clipperton Zone and elsewhere in abyssal seafloor areas. These nodules have a rather limited distribution of not more than a few square kilometres extent being restricted to the upper slopes of seamounts and intermontane basins between seamount peaks, generally in areas of fairly rugged topography. It is unlikely that sufficient of them exist in any one place to comprise a minesite as currently envisaged. This fact, coupled with the difficulties of the topographic setting make mining them under currently projected deep sea mining scenarios unlikely. One of the points in favour of the deeper water cobalt-rich nodules in the Cook Islands Exclusive Economic Zone is that the deposits are extensive and could be mined using the same techniques as have been developed for Clarion-Clipperton Zone nodules.

7

Cobalt–rich manganese crusts

Recent interest in the economic potential of manganese crusts lies principally in their cobalt content and to a lesser extent manganese and platinum. However, the observation of cobalt enrichment in shallow water seamount manganese nodules and crusts is not a new one. Menard (1964) drew attention to the general increase in the cobalt content of oceanic ferromanganese deposits with decreasing depth, while Mero (1965) and Cronan and Tooms (1969) pointed out that areas of cobalt enrichment in ferromanganese oxide deposits were centred on regions of elevated topography mainly in the western and southern Pacific. High cobalt values were recorded both in the Mid-Pacific Mountains and the Tuamotu Archipelago. Nevertheless, economic interest in these deposits was overshadowed by interest in the manganese nodules, and until lately their economic potential has been considered to be low.

Several factors have led to a reappraisal upwards of our perception of the economic potential of cobalt-rich manganese crusts. First, they tend to occur in shallow water areas, shallower by more than 2000 metres than most nodule deposits of economic interest, and sometimes within a few hundred metres of the sea surface (Figure 7.1). Second, they are thought to exhibit a high degree of coverage of exposed rock surfaces on the seafloor. Third, cobalt is a strategic metal of high market value present in low concentrations in most nickel and copper rich nodules, but present in maximum concentrations of up to 2.5% in some crusts. It finds a use in super alloys in high technology applications such as jet engines. Platinum is also relatively abundant in some crusts. Fourth, by their very occurrence on seamounts and, to a much lesser extent, on the submerged portions of volcanic islands (Figure 7.1), many cobalt-rich crusts fall within the EEZs of the adjacent island states and thus will be subject to the jurisdiction of those states rather than the International Seabed Authority.

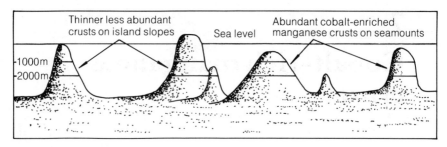

Fig. 7.1 Occurrence of cobalt-rich manganese crusts on seamounts and island slopes.

7.1 NATURE AND OCCURRENCE

Much of our present knowledge on the nature and occurrence of manganese crusts is from the Pacific. During a cruise in the Line Islands and Mid-Pacific Mountains region in 1981, Halbach *et al.* (1982) observed abundant manganese crusts on seamount slopes and summit plateaux. In the intervening valleys, however, no manganese deposits were detected. Manganese nodules were also present on the seamount plateaux in concentrations of up to 14 kg/m², and were most abundant in water depths of 1100–1900 m. These were the cobalt-rich seamount nodules referred to in the previous chapter and by Humphrey (1988). Encrustations were almost always present on seamount slopes between 1100 and 3000 m depth. The composition of the deposits showed considerable variations. Manganese varied from 15% to 31% and iron from 7% to 18%. The manganese/iron ratio lay between 1.0 and 3.4. No significant compositional differences between crusts and nodules at seamount stations were observed. Cobalt was found to be enriched up to about 2% in samples from summit regions (less than 1500 m depth), and averaged 0.8% in all samples recovered. The surfaces of seamount encrustations generally showed a knotty texture. However, crusts from summit regions sometimes had a smooth surface. Some crusts showed a fragmentation texture that is thought to have been caused by processes of dehydration during ageing of the deposit. Crust thicknesses ranged from less than 1 cm to more than 8 cm with average thickness of around 2.5 cm. In the western Pacific, thick encrustations are often characterized by two or more periods of growth. The lower part of the crust is sometimes followed by a phosphorite layer and above this a younger crust generation has grown. According to Halbach and Puteanus (1984) the younger generation of two generation crusts is, on average, richer in cobalt than the older crust.

Although the presence of two generations of crust has been clearly established in some areas, the results of a study by Clark *et al.* (1985) provide evidence for the occurrence of a third and older generation of crust in the Marshall Islands/Wake Island areas. The lower parts of these crusts may have been formed during a period of crust formation that preceded that over much of the remainder of the mid-Pacific area. More recently, Kang *et al.* (1990) have reported up to seven layers in crusts from the western Marshall Islands. Interestingly, however, some crusts in the Wake Island area up to 8 cm thick only show one crust generation. These contain 0.6–0.8% cobalt and occupy the tops of guyots and plateaux (Date, personal communication, 1989).

The thickness of crusts appears to be controlled, among other things, by the length of time that the growth has taken place and the rate at which the ferromanganese oxides have accumulated. This in turn partly depends upon oceanographic and geochemical conditions at the site of formation. Geological stability is required for the growth of very thick crusts, because slumping and erosion on seamounts can destroy or abrade crusts thus reducing their thickness (Hein *et al.*, 1985b). According to Manheim (1986), it appears that significant crust coverage occurs mainly on seamounts 60–80 million years old or older. Sedimentation must be prevented in order for crusts to form, and for this reason sedimented slopes of seamounts appear to be depleted in crusts relative to non-sedimented areas.

Work by Segl *et al.* (1984), and Halbach *et al.* (1983), has shown that the growth rate of crusts ranges from about 2.7 mm/million years for younger crusts to 4.8 mm/million years for older crusts, which is generally less than that for manganese nodules (Figure 7.2). The slower the formation of the crusts, the higher their cobalt content. Maximum cobalt contents of crusts of around 2% are believed to occur only in the most slowly growing crusts (Manheim, 1986).

Examination of the distribution of crusts in relation to latitude has shown that the maximum thickness and occurrence of crusts occur within 5–15° of the equator. In a study of the abundance and thickness of crusts from north to south in the entire Pacific Ocean, Humphrey (1988) has shown a clear increase in crust abundance as the equatorial zone is approached, but little variation in abundance within that zone. Cobalt and nickel abundances also increase equatorwards (Figure 7.3). Hodkinson and Cronan (1991) have substantiated and amplified these conclusions. Notwithstanding

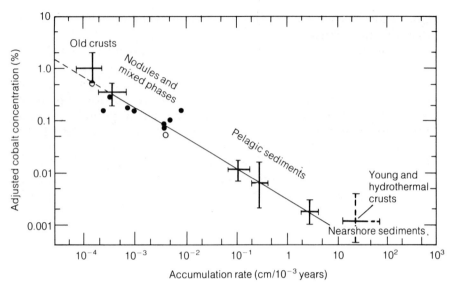

Fig. 7.2 Rate of accumulation vs. cobalt concentrations in cobalt-rich crusts, manganese nodules, hydrothermal crusts and nearshore and pelagic sediments (bars represent ranges) (from Manheim, 1986).

these observations, the localized distribution and thickness of crusts within any one seamount group can be quite variable, and variable between adjacent groups (Keating, 1989). The presence of modern reefs, especially large atolls, is unfavourable for the formation of crusts because large quantities of reefal debris migrate down the flanks of the island slopes, effectively inhibiting the formation of abundant cobalt-rich crusts. Within the upper 500–800 m of reef capped features, only thin stains of manganese oxide are generally found.

7.2 MINERALOGY AND COMPOSITIONAL VARIABILITY

The main manganese oxide phase to be identified in ferromanganese oxide encrustations is the δMnO_2 phase which is common in hydrogenetic nodules. This has X-ray reflections of 1.4 Å and 2.4 Å. The 9.7 Å reflecting mineral 10 Å manganite occurs rarely (Colley *et al.*, 1979) and occasionally other minerals have been noted. Amorphous material is also present.

Several elements in addition to manganese, iron and cobalt, mentioned earlier, show considerable ranges. Nickel values average

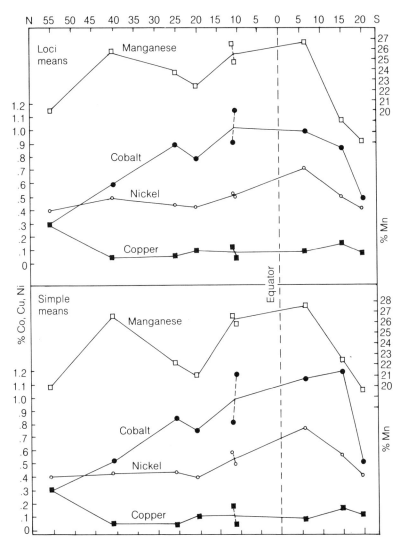

Fig. 7.3 Variability in the composition of cobalt-rich manganese crusts from north to south in the Pacific Ocean (from Humphrey, 1988).

between about 0.7% and 0.8% in the crusts but copper is generally very low, only around 0.05%. Cobalt, nickel and the manganese/ iron ratio show an inverse relationship to water depth (Figure 7.4). Copper tends to increase with water depth. Lead, zinc and molybdenum also tend to increase with decreasing depth. Maximum cobalt values of around 2.5% have been found in crusts on the SP Lee

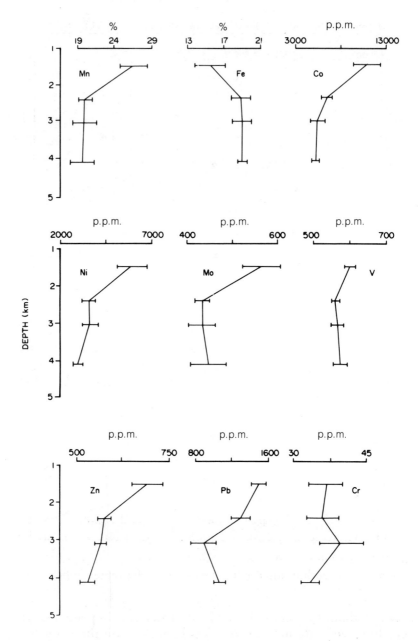

Fig. 7.4 Variations of elements with depth in manganese crusts from the Line Islands region (from Aplin, 1983).

Seamount at 8°N latitude in the central Pacific, and cobalt values of about 1% have been obtained from crusts in the Marshall Islands in the western Pacific (Hein *et al.*, 1988b). Very high values of more than 2% cobalt have been obtained from samples above 1000 m water depth in the southern Line Islands-French Polynesia area (Pichocki and Hoffert, 1987).

Cobalt enrichment in ferromanganese oxide encrustations and seamount nodules is not restricted to the Pacific Ocean. In the Atlantic, for example, on the Sierra Leone Rise and on the east flank of the Mid-Atlantic Ridge, crusts contain cobalt concentrations approaching 1% (Goddard *et al.*, 1987). However, to date, no crusts with the very high cobalt concentrations found in the central Pacific area have been recovered from the Atlantic. Cobalt-rich encrustations also occur in the Indian Ocean. Cronan and Tooms (1969) reported elevated cobalt concentrations in crusts from the south-western Indian Ocean, and Colley *et al.* (1979) have noted compositional trends with depth in crusts from the north-western Indian Ocean.

There is a strong negative correlation with latitude for manganese cobalt, nickel, molybdenum and cadmium in the outer layers of Pacific crusts. According to Hein *et al.* (1985b), these metals increase with proximity to the equator from the EEZ of Hawaii. Similar trends from north to south on an ocean-wide basis have been noted by Humphrey (1988) and Hodkinson and Cronan (1991).

In addition to cobalt in the crusts, a metal that has also been suggested to be of economic value in them is platinum. Halbach *et al.* (1984) drew attention to the presence of significant amounts of platinum within some crust samples and reported anomalously high platinum contents in Mid-Pacific Mountains crusts. More recent analyses from the Line Islands have produced values of up to 1.3 ppm platinum. Most recently still, crust analyses from the Hawaii EEZ have produced platinum values of up to almost 2 ppm (Wiltshire, 1990). Halbach *et al.* (1984) concluded that seamount crusts contain about 80 times more platinum than the upper continental crust. Maximum platinum values were found to occur in samples from water depths of less than 1250 m and within the two tier crusts the older crusts are significantly richer in platinum than the younger crusts. However, in a more recent study on about 70 crust samples from the Hawaiian and Johnston Island areas, Wiltshire (1990) found no relationship of platinum to depth. In comparison to the crusts, Bushvelt platinum ores in South Africa contain about 7–10 ppm total platinum group elements, and those in the Sudbury Complex contain about 3.3 ppm platinum.

Conditions under which cobalt enrichment in manganese crusts takes place have been the subject of a considerable amount of research. Halbach *et al.* (1982) noted a significant positive correlation between cobalt and manganese in the crusts that they examined, implying the cobalt abundance is controlled by the manganese content of the colloidal ferromanganese oxide particles that have been deposited from the bottom seawater. Cronan (1977) suggested that a reason for cobalt enrichment in such deposits was their slow growth rate and concomitant increased scavenging ability. Halbach *et al.* (1982) support this conclusion and have gone further in suggesting that only cobalt among the important minor metals in oceanic ferromanganese deposits increases in concentration with decreasing growth rates of the deposits. These conclusions have been amplified in a suite of crusts from the Line Islands by Aplin and Cronan (1985a). They consider that the dominant controls on the minor element composition of the crusts are the surface properties and crystal chemistry of manganese oxide and iron oxide flocculates. Since both manganese and iron oxides are known to be efficient scavengers of Mn^{2+}, it can be envisaged that excess dissolved manganese is being adsorbed on to the manganese and iron oxide flocs and therefore into the crusts. This excess manganese in turn scavenges cobalt from seawater, thereby enriching it in the crusts also.

Klinkhammer and Bender (1980) have drawn attention to an enrichment of dissolved manganese in water depths between 900 m and 1300 m in the central Pacific, closely associated with the dissolved oxygen minimum. This is the result of a variety of processes (Chester, 1990). From this zone of maximum dissolved manganese concentration, there is a diffusion flux of manganese into deeper water layers with its subsequent precipitation on available surfaces such as the upper slopes of seamounts where it contributes to crust growth (Halbach *et al.*, 1989), and scavenges cobalt.

The dissolved oxygen minimum zone is best developed in the northern equatorial Pacific (Figure 1.2) and on the basis of the above model one might expect manganese crusts to be better developed in the northern equatorial Pacific than in the southern Pacific. This indeed seems to be the case with crusts of up to 8 cm or more thickness in the Marshall Islands and the Wake Island area while crusts in the south equatorial Pacific (see case study below) tend to be much thinner. Possibly the intensity of the oxygen minimum zone is affected by the degree of biological productivity in the overlying waters. On this basis, it would be tempting to link all cobalt enrichments in crusts to the equatorial high productivity belt by its influence on the dissolved oxygen minimum zone. However, values of cobalt greater than 1% have also been found in areas of lower productivity such as the western part of the Hawaiian Ridge and French Polynesia. Some cobalt enrichments in old crusts to the north of the equator may be relict inasmuch as the substrates which they coat will have passed through the equatorial zone of high productivity in the past. Cobalt enrichments in the Pacific just to the south of the equatorial zone of high productivity may also be related to the dissolved oxygen minimum zone which is developed there as far south as 15°S (Figure 1.2). However, the presence of a regional dissolved oxygen minimum may not be necessary for cobalt enrichment in crusts if horizontal advection occurs or if there is local obstructional upwelling around seamounts and ridges leading to elevated productivity in the surface waters, and a localized oxygen minimum zone under them.

Such obvious relationships between the nature of crusts and modern oceanographic conditions suggest that many crusts may be much younger than their substrate ages. This would be in accord with discrepancies in crust thickness relative to that expected on the basis of crust accumulation rate data and known substrate ages. Many crusts are too thin to have been accumulating con-tinuously since their substrates were formed. Several factors could be responsible for this such as slumping, crust abrasion, burial and local volcanic activity. Indeed, it would be remarkable if seamount slopes had remained sufficiently stable for continuous crust growth during the many millions of years that have often passed between substrate initiation and the present day.

7.3 CASE HISTORIES

Two contrasting examples of cobalt-rich crust regions are described, one from north of the equator and one from the south.

7.3.1 Cobalt rich manganese crusts in the EEZs of Hawaii and Johnston Island (Figure 7.5)

Cobalt-rich crust data from the Hawaiian EEZ and Johnston Island EEZ have been provided by Helsley *et al.* (1985), Halbach and Puteanus (1985) and the State of Hawaii Marine Mining Programme (1987).

Crust thicknesses have been estimated to vary from stains to 12 cm with the greater thicknesses occurring on older seamounts away from the main hot spot axis of the Hawaiian Chain (Helsley *et al.*,

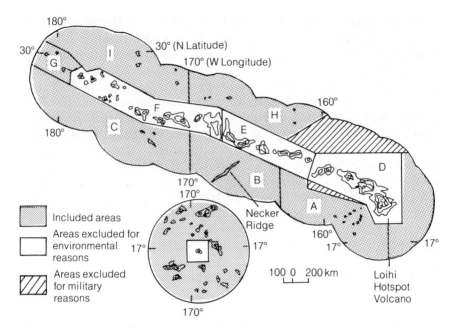

Fig. 7.5 Areas of cobalt-rich crust investigations in the Hawaiian and Johnston Island EEZs (from Helsley *et al.*, 1985) and Loihi Hotspot Volcano.

1985; State of Hawaii Marine Mining Programme, 1987). Thicker crusts occur on solid attached substrates than on loose cobbles and boulders. Crust thicknesses exhibit no consistent relationship to substrate types except that those on carbonate substrates tend to be thinner than those on other substrate types. Basalt is the main substrate present. Crust densities average about 1.95 wet and 1.33 dry.

The average cobalt content of the Hawaiian deposits is 0.75%, with the average value for on-axis Hawaiian Archipelago sites being 10% greater than for off-axis sites. Manganese values for the two groups are roughly the same. Actual average values are given in Table 7.1. Statistical analysis of the geochemical data from the Hawaiian Archipelago suggests that cobalt, manganese and nickel decrease with depth and that iron and copper concentrations increase with depth.

Some samples recovered from off-axis seamounts show multiple growth episodes. These samples often contain significant amounts of phosphorite infilling. The innermost ferromanganese crust probably represents early in-situ crust formation at the highest elevations,

Table 7.1. Composition of Hawaiian and Johnston Island cobalt-rich crusts (after State of Hawaii Marine Mining Programme, 1987)

Sub-area	Site	ND/NS	Average crust composition Co	Ni	Mn	Fe
D	AX10	10	0.72	0.37	20.6	19.1
	AX11	1	0.04	0.08	11.6	10.8
			0.38	0.22	16.0	15.0
E	AX6	8	0.58	0.34	18.9	18.5
	AX8	3	0.98	0.35	20.9	15.6
	AX9	3	1.11	0.41	24.6	16.4
			0.89	0.37	21.5	16.8
F	AX3	4	0.59	0.31	22.3	12.6
	AX4	6	0.94	0.40	21.6	16.4
	AX5	1	0.31	0.19	6.8	17.8
	AX7	2	0.78	0.56	27.6	9.9
			0.65	0.36	19.6	14.2
G	AX1	6	0.71	0.37	20.9	17.3
	AX2	5	1.07	0.64	27.4	13.1
			0.89	0.50	24.1	15.2
A	OF4	3	0.42	0.28	17.8	20.9
	CROSS[a]	55	0.53	0.30	18.2	18.2
			0.47	0.29	18.0	19.6
B	OF5	8	0.55	0.28	19.3	20.0
C	OF6[b]	6	0.63	0.33	18.6	18.7
H	OF2	6	0.58	0.33	19.3	18.9
	OF3	5	0.72	0.33	21.0	19.0
			0.65	0.33	20.2	18.9
I	OF1	6	0.95	0.48	25.3	15.3
Johnston Island EEZ		–	0.95	0.67	25.7	13.1
On-Axis Average		11	0.71	0.37	20.3	15.2
Off-Axis Average		6	0.63	0.33	18.6	18.7
Archipelago Average		17	0.68	0.36	19.7	16.4

AX – On-axis site

OF – Off-axis site

ND – Number of dredge-haul averages used to compute the site averages

NS – Number of site averages used to compute averages for on- and off-axis and total study area

Notes: (a) Data from September 1985, University of Hawaii expedition to Cross Seamount, personal communication from A. Malahoff, principal investivgator

(b) Values are averages of off-axis sites. Replicates are for number of sites (6)

Sub areas are shown in Figure 7.5.

whereas subsequent layers may have grown on the rock debris at greater depths after its downslope movement (Helsley *et al.*, 1985).

Average slope angles vary from 12° to 20° on off-axis seamounts, from 8° to 12° on on-axis seamounts, and from 5° to 8° on the Necker Ridge. Because of its age and minimal slope angle, the Necker Ridge is likely to have a greater resource potential for ferromanganese crusts than the Hawaiian Chain. It also contains thicker crusts (Helsley *et al.*, 1985).

7.3.2 Cobalt–rich manganese crusts in south equatorial Pacific EEZs

One area that has been looked at in some detail for cobalt-rich crusts is the Line Islands. Using crust analyses from most of the length of the Line Islands, Aplin and Cronan (1985a) carried out an analysis of variance procedure to examine the effect of various environmental parameters on crust composition. More recently, Hodkinson and Cronan (1991) have applied the same techniques as were used in the Line Islands to a much larger dataset from the Central Pacific (Figure 7.6). Analysis of variance techniques, together with mean data for differing latitudinal provinces were used to assess crust compositional variability with water depth, longitude and latitude. No significant relationships were found between measured crust parameters and longitude.

As would be expected, crust composition showed significant variations with depth for all elements studied. Manganese, cobalt and nickel all decreased with increasing water depth whereas iron and copper increased with increasing water depth. These results confirmed the findings of Aplin and Cronan (1985a), and in addition show that the composition/depth trends outlined by these workers for the Line Islands are applicable to crusts throughout a much wider area of the central Pacific.

In respect to crust variation with latitude, variations in crust composition have been noted previously with the enrichment of manganese and associated elements occurring with increasing proximity to the equator (Humphrey, 1988). In order to assess the variation in crust chemistry with latitude more closely in the equatorial region, Hodkinson and Cronan (1991) plotted elements against latitude for 333 crust samples between 30°N and 25°S. (Figure 7.7). For each 500 m depth interval, the plot showed almost identical trends to those for the total dataset, with only the absolute concentrations varying between each depth interval. Manganese, cobalt and nickel reach maximum values between 3°S and 14°N (Figure 7.7).

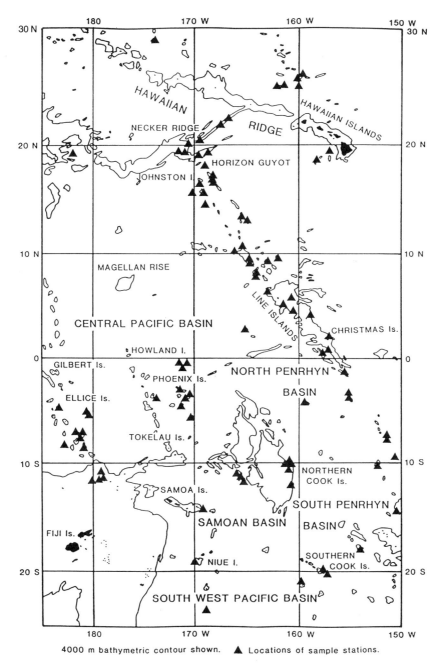

Fig. 7.6 Distribution of crust samples studied from the South Pacific (from Hodkinson and Cronan, 1991).

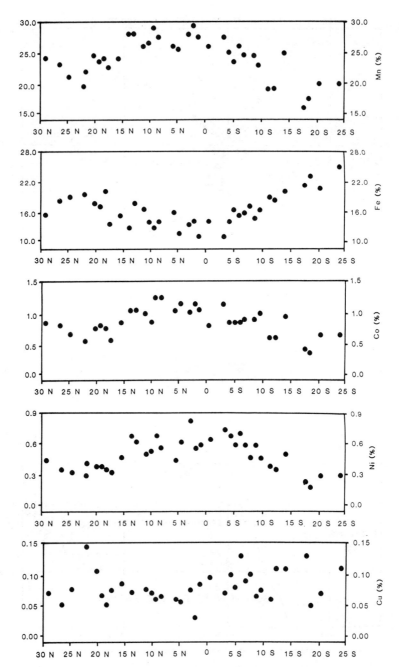

Fig. 7.7 Latitudinal variations in crust composition in the Central Pacific (from Hodkinson and Cronan, 1991).

Latitudinal variations in crust thickness have been found by Hodkinson and Cronan (1991) to be more asymmetric about the equator than those of chemical composition. Minimum thicknesses occur around 20°S, rising sharply north of 10°S. To the north of this, average crust thickness shows random variations before starting to decrease at about 15°N.

7.4 RESOURCE POTENTIAL OF EEZ COBALT-RICH CRUSTS

Manheim (1986) pointed out that concerns about cobalt came into prominence in 1978, when one of its main sources, Zaire, was invaded. The price of the metal increased as a result of this, although supply was not substantially reduced. However, this did point to the vulnerability of much of the world's cobalt supply. According to Manheim (1986), enrichments of cobalt to levels greater than 0.1% on land are confined to only a few major deposits such as those in the copper belt of central Africa. In relatively few areas do cobalt minerals actually dominate ore deposits. The large African deposits have cobalt as a minor constituent, with concentrations of only up to about 0.4% in the Zairan deposits and values nearer 0.1% in deposits from the Zambia copper belt, Cuba and Australia.

In view of these observations, it is not surprising that the possibility of mining cobalt-rich crusts has received much attention. US Bureau of Mines studies reported in Wiltshire (1990) suggest that by 2005, price rises in cobalt will make cobalt-rich crust mining economic. It has been estimated that the output of a cobalt crust mining operation of capacity 1 million wet tonnes per year could potentially meet a significant part of US cobalt demand, about one-quarter of the manganese demand and small fractions of the demand for other metals such as nickel, copper, platinum, vanadium and molybdenum (Manheim, 1986). One area in the United States Exclusive Economic Zone where crusts occur in abundance is the Blake Plateau, where, together with nodules, the deposits cover large areas. As mentioned, these deposits have been suggested as being of use for catalyst purposes. Some metals might also be extracted from them, even though cobalt, nickel and manganese generally tend to be present in lower concentrations than in both nodules in the deep ocean and crusts in the Exclusive Economic Zones of the Pacific.

Hein *et al.* (1987) developed 11 criteria for the location and possible exploitation of cobalt-rich crusts. Exploration criteria for locating the crusts included large seamounts shallower than 1500 m, substrate rocks older than 20 million years, areas of strong current activity, absence of atolls or reefs on top of the seamounts, a well developed oxygen minimum zone, lack of slumping and mass wastage on seamount slopes, lack of local volcanic activity, and lack of significant terrigenous detritus. Exploitation criteria included more than 0.8% cobalt, more than 4 cm thick crusts, and relatively low topographic variations.

According to Clark *et al.* (1985), in resource assessments for crusts, it is assumed that commercial concentrations of the deposits would be confined to slopes and plateaux in water depths between 800 and 2400 m; be most common in areas that are older than 25 million years and with low sediment cover. In assessing the resource potential of cobalt-rich crusts in parts of the western Pacific, Clark *et al.* (1985) considered the following data to be of importance: the area of a nation's Exclusive Economic Zone, the 'permissive area' between 800 and 2400 m where crusts are most likely to occur, the portion of the permissive area expected to contain potentially commercial crust concentrations, the composition and thickness of the crusts and their wet and dry density.

Using these parameters, Clark *et al.* (1985) carried out a resource assessment of cobalt-rich crusts in the Exclusive Economic Zones of United States Trust Territories of the western Pacific (Table 7.2), where both single crust and two crust generations have been reported.

Based on the data presented in Table 7.2, Clark *et al.* (1985) ranked the US Trust and Affiliated Territories in decreasing order in terms of potential for cobalt-rich crusts.

> Federated States of Micronesia
> Marshall Islands
> Commonwealth of Marianas
> Kingman–Palmyra
> Johnston Island
> Wake
> Jarvis
> Belau–Palau
> Guam
> Howland–Baker
> Samoa

This listing mainly reflects the total permissive area of each of the Territories. Within the area of the Federated States of Micronesia, Clark *et al.* (1985) conclude that the permissive area for single crust occurrences is 176 809 km^2 and for two-crust generation occurrences is 92 161 km^2. The largest permissive area for the occurrence of both

Table 7.2. Resource assessment of cobalt-rich crusts in parts of the western Pacific (from Clark *et al.*, 1985)

Territory	Resource potential			
	Cobalt	Nickel (t × 10⁶)	Manganese	Platinum (oz × 10⁶)
Belau/Palau	0.55	0.31	15.5	0.68
Guam	0.55	0.31	15.5	0.68
Howland–Baker	0.19	0.11	5.5	0.48
Jarvis	0.06	0.03	1.6	0.15
Johnston	1.38	0.69	41.6	3.5
Kingman–Palmyra	3.38	1.52	76.1	5.7
Marshall Islands	10.55	5.49	281.3	21.5
Micronesia	17.76	9.96	496.0	34.7
Northern Marianas	3.60	1.97	100.2	7.7
Samoa	0.03	0.01	0.8	0.04
Wake	0.98	0.51	26.8	2.0

Note: The above are estimates of in-place resources and as such do not indicate either potential recoverability or mineable quantities.

crust generations is a broad plateau area which may be sediment covered, thereby possibly reducing the estimated crust potential.

The permissive areas (single and double crust generation) of the Commonwealth of the Marianas are approximately equally distributed (24 881 km² versus 20 388 km²). However, Clark *et al.* (1985) point out that the ages of seamounts in the older portion of the area are poorly known and could be much younger than inferred. If so, the resource potential of the area would be considerably reduced.

An estimation of the resource potential of cobalt-rich crusts in Exclusive Economic Zones in the south equatorial Pacific has been carried out by Cronan *et al.* (1989) and Cronan and Hodkinson (1989, 1990) using the same approach as was used by Clark *et al.* (1985) in the Exclusive Economic Zones of the United States Trust Territories. The nature of the crusts in these areas has been described on. p. 126 and the same data, coupled with the additional parameters used by Clark *et al.* (1985), were used to calculate the crust resource potential of those areas.

On the basis of the data available, crusts from the Exclusive Economic Zones of the Cook and central Line Islands, for example, show many features in common with central Pacific crusts in general. Their average cobalt content is 0.68%, slightly less than the central Pacific average of 0.79% reported by Hein *et al.* (1990). Their average thickness is 1.4 cm with a range of thicknesses from 2 mm to

4 cm. Using the assumptions of Clark *et al.* (1985) with minor modifications, the permissive area of the Cook Islands was estimated to be nearly 25 000 km^2 with a crust potential of almost 140 million tonnes containing about 930 000 tonnes of cobalt, 630 000 tonnes of nickel, 109 000 tonnes of copper and about 60 tonnes of platinum.

On the basis of this calculation, it is interesting to note that cobalt tonnages in the crusts of the Cook Islands Exclusive Economic Zone are much lower than cobalt tonnages in the nodules there (p. 112). This calls into serious question the possibility that cobalt–rich crusts represent a greater cobalt resource than nodules in the central Pacific. On the basis of the figures presented by Cronan *et al.* (1989), nodules would appear to be a much greater cobalt resource than crusts in the Cook Islands Exclusive Economic Zone.

Similar calculations of crust metal tonnages to those done in the Cook Islands Exclusive Economic Zone have been made for the Phoenix Islands and Ellice Islands (Tuvalu) EEZs by Cronan and Hodkinson (1989, 1990). In the Phoenix Islands EEZs, the calculations show that there are approximately 125 000 tonnes of combined nickel and cobalt in crusts and in the Ellice Island EEZ approximately 1.2 million tonnes. In the case of the Phoenix Islands EEZ, this is substantially less than the amounts of those metals present in manganese nodules there (p. 113) and in the case of the Ellice Islands EEZ (p. 113) the two values are approximately similar.

It would seem clear from these data that the amounts of potentially mineable metals present in cobalt–rich crusts within the Exclusive Economic Zones situated in the south equatorial Pacific are far less than those potentially available in manganese nodules in these areas, and therefore the nodules represent a greater resource than the crusts there, even for cobalt. Whether these conclusions can be extrapolated to other areas is conjectural. Comparing the south equatorial Pacific Exclusive Economic Zones with those, for example, of the US Trust Territories, it is clear that nodules are much more abundant in the former than they are in the latter, and thus cobalt–rich crusts are likely to offer a greater resource potential than nodules in the Trust Territories. However, it could also be argued that as the market for cobalt is small, cobalt extracted from nodules would be sufficient to meet world cobalt demand, and there would be no need of crust mining for cobalt. Wiltshire (1990) argues that the platinum in crusts may make the difference between an economic and non-economic recovery operation. Possibly, the only economically viable crust deposits would be those of great thickness, say over 10 cm, very rich in cobalt and platinum, and containing a phosphate substrate or interlayer which could also be recovered and sold.

8

Hydrothermal Mineral Deposits

Hydrothermal mineral deposits, including potentially economic sulphide minerals, are not primarily Exclusive Economic Zone deposits, although many occurrences of them have been reported in Exclusive Economic Zones. Polymetallic sulphides and their associated metalliferous sediments consist of sulphides of iron, copper and zinc, oxides and silicates of iron, and oxides of manganese, formed by hydrothermal processes associated with sea-floor volcanic activity. They are often associated with barite, silica and anhydrite. Most of the occurrences that have been reported have been on mid-ocean ridge spreading centres, including the Red Sea, and to a lesser extent at convergent plate margins (Figure 8.1), but they also occur on hot spot volcanoes in intra-plate settings. Most of the deposits on mid-ocean ridges occur in the International Seabed Area but are very similar to others reported in Exclusive Economic Zones. By contrast, most of the convergent plate margin deposits and, of course, the Red Sea deposits, fall within Exclusive Economic Zones.

8.1 MID-OCEAN RIDGE HYDROTHERMAL DEPOSITS

8.1.1 Nature and origin

Mid-ocean ridge hydrothermal deposits are formed as a result of seawater entering sub-seafloor rocks through cracks and fissures, leaching metals from the rocks and becoming transformed into a mineralizing solution which rises back to the sea floor to precipitate leached metals as hydrothermal mineral deposits. The driving force for these reactions is a heat source supplied at mid-ocean ridges by magma injected into the crust. As oceanic plates diverge at mid-ocean ridges, magma rises and becomes injected into crustal rocks in the form of dykes subsurface, and extrudes as lava flows on the

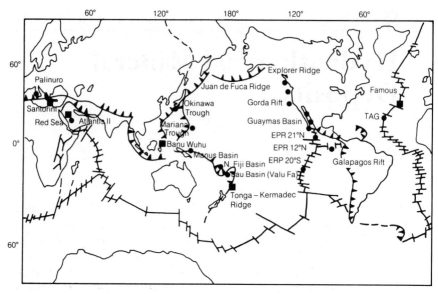

EPR = East Pacific Rise.

— Divergent plate boundaries and rift zones; ▲ collision and subduction zones; ■ oxidic metalliferous deposits; ● black smokers and polymetallic sulphide deposits;

Fig. 8.1 Examples of submarine hydrothermal mineralization at divergent and convergent plate margins (modified from Cronan, 1985).

surface. As new sea floor moves away from the spreading axis of the ridge, it cracks and fractures, allowing seawater to enter. Such seawater penetration may occur in a relatively wide downwelling zone to as deep as 2–3 km below the seafloor. During this process, it becomes heated and leaches metals from the basalt through which it passes. According to Lange (1985a), several reactions can take place which drastically alter the composition of the seawater and which have been reproduced in laboratory experiments. First, there is an increase in the acidity of the seawater by incorporation of Mg^{2+} ions and OH^- ions into magnesium hydroxysilicate phases, accompanied by the production of hydrogen ions. Second, cation exchange reactions of hydrogen ions with the minerals in the basalt leach transition metals such as iron, manganese, copper and zinc among others from them. Third, there is a reaction of calcium with sulphate ions to form anhydrite. Fourth, sulphide ions react with iron to form pyrite; sulphides of Cu + Zn also precipitate. Fifth, enrichment of sodium and chlorine in the solutions occurs because of their concentration during hydration reactions. The sulphur which takes part in these reactions is thought to be derived both from the reduction of seawater sulphate and the dissolution of sulphide minerals of magmatic origin in the basalts being leached.

Eventually, the hot, acid, metal-rich hydrothermal solutions rise back towards the sea floor with the metals largely transported as chloride complexes. Should they encounter significant amounts of downwelling cold seawater during their ascent, the bulk of their metals will be precipitated as a stockwork of sulphide minerals within the upper part of the oceanic crust. On the other hand, if the conduits through which the hydrothermal solutions pass remain sealed to seawater, an undiluted hydrothermal solution will discharge on the sea floor as a smoker, a suspension of finely divided hydrothermal precipitates from which hydrothermal mineral deposits will precipitate.

According to Edmond *et al.* (1979), hydrothermal discharges on mid-ocean ridges are variable in composition depending on the extent to which they have become mixed with seawater during their ascent and sub-sea floor fractionation has taken place. Some are quite diluted and discharge at temperatures of less than 40°C, while others are much more concentrated and have temperatures up to 400°C. The latter form two types of smokers, black smokers and white smokers. Black smokers consist largely of fine-grained precipitates of sulphide minerals discharging at high temperatures. By contrast, white smokers are of lower temperature and contain finely divided precipitates of sulphate minerals and silica. Some of the minerals in the smokers precipitate as chimneys in the immediate vicinity of the vents but much gets up two or three hundred metres into the water column to form a hydrothermal plume which then disperses away from the site of hydrothermal discharge under the influence of prevailing ocean currents. Within the plume, manganese and iron are oxidized and, together with sulphide particles, fall out of the plume and accumulate as metalliferous sediments over wide areas of the crest and flanks of the ridges (Figure 1.3). The presence of these fallout products from the plume have been detected up to almost 1000 km away from the spreading axis of the East Pacific Rise in the South Pacific (Edmond *et al.*, 1982) and on a more localized scale can be very useful in geochemical exploration for submarine hydrothermal mineral deposits by helping to pinpoint the sites of hydrothermal discharge (see Cronan, 1985 for a review).

Not all seafloor hydrothermal mineral deposits precipitate from solutions which have simply formed as a result of basalt-seawater interaction. Sometimes, fractionation processes occur between leaching of the seafloor basalts and final precipitation of the leached metals as mineral deposits, which can lead to a considerable diversity of deposit types (Backer and Lange, 1987). If the hydrothermal solutions have to pass through sediments before discharge on to the sea floor, reactions with the sediment can take place which lead to the discharge of solutions different from those which would have discharged through bare rock. For example, in the hydrothermal mounds of the Galapagos Rift, reactions between hydrothermal solutions and pelagic sediments have resulted in the sediments being replaced by an iron-rich smectite overlain by manganese oxides at the sediment water interface (Moorby and Cronan, 1983). In the Guaymas Basin, Gulf of California, high temperature hydrothermal solutions have reacted with sediments, resulting in the dissolution of calcium carbonate and siliceous organisms, and the precipitation of a variety of new hydrothermal mineral phases. The composition of these hydrothermal deposits, and those of the vent solution and altered sediments, suggest that hydrothermal fluids form in the basaltic basement complex underlying the Guaymas Basin and then interact with organic and carbonate rich material on ascent, followed by near surface mixing with ambient seawater (Koski *et al.*, 1985). Petroleum has formed from the transformation of organic material in the sediments (Simoneit and Lonsdale, 1982). According to Koski et al. (1988), on the sediment covered floor of the Escanaba Trough, Gorda Ridge, in the US EEZ, the mineralogy, metal content, sulphur isotope composition and hydrocarbon content of massive sulphides reflect an extensive interaction between underlying tur-bidite sediments and hydrothermal fluids. Sulphide mounds rich in pyrrhotite are thought to have formed under gentle chemical and temperature gradients related to diffuse low velocity flow of hydrothermal solutions through sediments. Such processes have resulted in the formation of around one million tonnes of mound and ridge-like deposits, stratified on the sediment surface (Koski, 1989). The pyrrhotite-rich massive sulphides contain up to 20% copper. Stannite is also present indicating that the hydrothermal fluids are able to transport tin. Gold is present in the massive sulphides in amounts up to 10.1 ppm, and averaging 3.2 ppm. These deposits in the Escanaba Trough are considered by Koski (1989) to be analogous to Besshi-type deposits, sediment hosted stratified sulphide ore deposits found in Japan.

Within hydrothermal brine pools in the Atlantis II Deep, Red Sea, (see case study below), a restricted environment may be maintained for long periods as a result of the stratification of saline waters, resulting in the selective precipitation of metals from hydrother-mal solutions as the physico-chemical nature of the depositional environment changes (Cronan, 1980). This leads to the early precipitation of sulphide minerals, the later precipitation of iron silicates and iron oxides, and lastly the precipitation of manganese oxides most of which escape from the main brine pools to precipitate in marginal deeps or as a halo in sediments around the Deep (Bignell *et al.*, 1976a) (Figure 8.2). According to Backer and Lange (1987), there is even some compositional differentiation among base metal sulphides *alone* with increasing distance from the Atlantis II Deep hydrothermal discharge centre.

As mentioned above, the hydrothermal solutions which discharge on to the sea floor precipitate a portion of their metals in the form of chimneys. These chimneys are highly variable, both in structure and mineralogical composition (Figure 8.3). According to Lange (1985a), chimney growth commences with a rapid deposition of a highly permeable wall composed of anhydrite and minor admixtures of wurtzite and pyrite. Mixing of seawater and hydrothermal fluid through this wall causes further precipitation of anhydrite, wurtzite and pyrite, reducing the permeability of the wall. This results in the temperature of the hydrothermal fluid inside the chimney rising, leading to sulphide deposition in the chimney interior. Once the majority of the openings from the chimney interior to the chimney exterior are sealed by hydrothermal precipitates, a high temperature hydrothermal assemblage develops in the chimney interior. Wurtzite and sphalerite are the main minerals in the interiors of zinc rich chimneys, chalcopyrite is the dominant mineral in the interiors of copper-rich chimneys. The temperatures of hydrothermal fluids leaving zinc-rich chimneys are generally below 300°C whereas fluids leaving copper-rich chimneys are between 300°C and 350°C. As the temperature of the chimney exteriors drops, alteration reactions between seawater and mineral assemblages in the chimney walls produce new mineral phases. According to Haymon and Kastner (1981), anhydrite dissolves at the outer edge of chimneys at temperatures below 130°C. Alteration products such as limonite form. A hard gossan of these alteration products is commonly observed on the exteriors of chimneys, and this serves to protect the chimney interiors from further alteration, thereby preserving them on the sea floor for considerable periods of time. However, even though alteration may be slowed, it probably never completely stops. For example, totally altered chimneys consisting of almost pure iron oxyhydroxides with minor opaline silica have been observed at an extinct hydrothermal site on the EPR by Hekinian *et al.* (1980).

Chimney growth can be quite rapid. Submersible observations have recorded growth rates of as much as 30 cm per day (Goldfarb *et al.*, 1983) in the initial stages of chimney growth and 8 cm per day

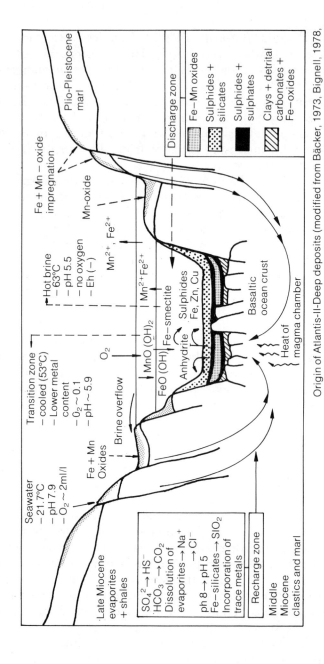

Fig. 8.2 Mineral forming processes in the Atlantis II Deep, Red Sea (from Lange, 1985b).

Origin of Atlantis-II-Deep deposits (modified from Bäcker, 1973, Bignell, 1978, Cole, 1983).

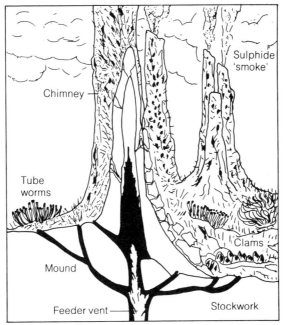

Chimney

Sulphide 'smoke'

Tube worms

Clams

Mound

Feeder vent

Stockwork

Hydrothermal vents occur along the floor of the axial valley of actively spreading ridge crests.

Fig. 8.3 Idealized cross-section of black smoker chimney interior growing on a mound of collapsed chimney material (modified from McGregor and Lockwood, 1985).

(Hekinian *et al.*, 1983) in older chimneys. The chimneys can grow to heights of as much as 20 m or more, after which they topple over to form a mound of broken chimney rubble. The intergrowth of adjacent chimneys can also contribute to mound formation (Figure 8.3).

Mounds are a characteristic feature of chimney sites. According to Speiss *et al.* (1980), sulphide deposits on the East Pacific Rise at 20°50′N consist of broad basal mounds up to 5 m high and 450 m in diameter. These are capped by active and inactive chimney structures up to a further 5 m high. However, sulphide edifices up to 26 m in height have been observed on the EPR at 12°50′N and up to 30 m high elsewhere. The surfaces of the mounds are covered by partially oxidized sulphides and fine grained sulphide mud. As mentioned, a portion of the mounds are probably accreted from the physical and chemical breakdown of the chimneys. However, some mounds have no chimneys on them. These must have formed in place by the convection of hydrothermal fluids beneath an impermeable crust (Goldfarb *et al.*, 1983). According to Rona (1984), a typical mound with chimneys on it contains approximately 1000 tonnes of metals.

8.1.2 Mineralogical and chemical compositional variability of mid-ocean ridge hydrothermal deposits

The composition of polymetallic sulphides on mid-ocean ridges, and to a lesser extent that of metalliferous sediments associated with them, is highly variable and depends upon several factors including temperature, the composition of the sub-seafloor rocks being leached, and the rock/water ratio at the site of leaching. Small amounts of gold, silver and other valuable metals have been found in some mid-ocean ridge sulphide deposits, and more recently in convergent plate margin varieties also.

According to Lange (1985a), much of the compositional variability in seafloor sulphides can be related to their mineralogical variability. He reports main components of the deposits as pyrite/marcazite, wurtzite/sphalerite, chalcopyrite/cubanite, anhydrite/barite, and amorphous silica. Lange attributes the variation of major constituents in the deposits to differences in the temperature of the exiting hydrothermal solutions. High temperature deposits are enriched in copper and zinc sulphides, low temperature deposits have lower concentrations of these sulphides and contain much amorphous silica and sometimes barite, while inactive deposits lack significant anhydrite and wurtzite and are coated by a gossan containing limonite and goethite accompanied by minor amounts of base metal sulphates.

Bulk analyses of ten polymetallic sulphide occurrences reported by Lange (1985a) confirm high values of zinc, copper and silver with major amounts (i.e. greater than 1%) of iron, zinc, silica and sulphur. The zinc–copper–iron ratio can be highly variable from deposit to deposit. Copper appears to be enriched in the southern East Pacific Rise and Galapagos Rift occurrences (Figure 8.1), compared with the more northerly deposits on the East Pacific Rise which are richer in zinc (Table 8.1). It has been reported that deposits on the East Pacific Rise at 21°N and on the Juan de Fuca Ridge are similar, being composed primarily of zinc, iron and sulphur with important minor concentrations of gold, arsenic, cadmium and germanium. However, as pointed out by Scott (1987), three dimensional sampling has not been carried out to any significant extent on mid-ocean ridge sulphide deposits and therefore their bulk composition is very imperfectly known. Most of our knowledge of the composition of these deposits comes from surface samples only, and thus reported compositional differences between them may not be representative of their overall compositional variability.

Notwithstanding this limitation, Scott (1987) reports significant differences in the composition of sulphide deposits hosted by basalt and sediments respectively. Low iron, copper and zinc contents, a high ratio of Pb/(Cu + Zn) and high values of barium and calcium, reported for Guaymas Basin samples compared to those from other sites can be explained by differences in the composition of their respective vent fluids, related in turn to different water-rock reactions in basalt and sediment. Scott reports that the Guaymas Basin sediments are rich in lead, calcium and barium, relative to other metals as a result of their being rich in carbonate and pelitic material, which further acts as a sink for iron, copper and zinc.

A compilation of amounts of gold and silver in mid-ocean ridge sulphides has recently been carried out by Hannington and Scott (1988). Values of from 1–16 ppm gold and 100–640 ppm silver have been found in sulphide samples from several spreading centres, including the Southern Explorer Ridge and Axial Seamount in the north-east Pacific, and the TAG hydrothermal field in the North Atlantic. The Southern Explorer Ridge samples may contain a total of 70 000–120 000 ounces of gold. The average gold content of 56 samples from these three deposits is 3.4 ppm. Average concentrations of silver in gold-rich samples are 140 ppm, but high values also occur in gold-poor samples. High concentrations of gold (>1 ppm) are generally associated with elevated lead, silver, arsenic and antimony concentrations in the deposits. Some of these variations could be artefacts of sampling, but, as pointed out by Scott (1987) and others, some variations could be real and related to compositional variation between the substrate rocks being leached by the hydrothermal solutions, selective scavenging processes or to selective precipitation processes. Hannington and Scott (1988) consider that the gold contents of submarine hydrothermal deposits are controlled by the chemistry and maturity of the hydrothermal systems, and propose a model whereby gold is preconcentrated to about 0.2–0.5 ppm in early precipitated high temperature Cu-Fe sulphides and is subsequently remobilized and concentrated (1–10 ppm) in late stage lower temperature precipitates in the outer

Table 8.1. Compositional variability of marine sulphide occurrences (from Lange, 1985a)

	JFR 47°57'N	JFR 46°00'N	JFR 44°40'N	EPR 20°54'N	EPR 20°50'N	EPR 12°50'N	EPR 18°31'S	EPR 21°26'S	GRZ 85°50'W	GRZ 85°54'W	Mean
Zn (%)	6.25	19.2	54.0	40.8	32.3	16.0	41.3	9.1	1.4	14.6	23.6
Cu	0.5	0.2	0.2	0.6	0.8	1.0	5.5	6.8	4.1	4.3	2.4
Pb	0.10	0.35	0.25	0.05	0.32	0.14	0.04	0.03	0.02	0.05	0.14
Fe		4.5	8.9	7.6	19.2	25.0	16.3	35.8	36.3	21.4	19.5
S			34.7	31.9	35.3	33.3	32.5	45.5	38.7	30.6	35.3
SO$_3$			<0.09		<2.5	0.5	<0.1	<0.1	0.53	0.34	<0.6
SiO$_2$	30.9	2.1	6.4	7.9		2.0	1.2	13.6	25.4	11.2	
Al$_2$O$_3$			0.11		0.39	0.40	0.38	0.25	0.85	0.30	0.4
CaO		0.21	<0.09		<1.8	0.35	0.01	0.01	0.14	0.04	<0.3
MgO			<0.05		<0.04	0.11	0.02	0.03	0.15	0.07	<0.06
Ba		14.6	0.06		0.23	0.03			0.17	0.04	2.1
Ag (g/t)	30	290	260	380	156	100	62	46	24	124	150
Au	0.08		0.13		0.17		0.20	0.30	0.20	0.93	
Cd	260	425	775	500	560	504	1100	280	60	230	470
Ge			195		65	57	152	56	10	20	80
Ga			<20		14	38	75	62	16	60	41
In						7	20	10	6		11
Bi			<0.2		<0.8	<5	<5	<5	<5	<5	<5

Sb			27		37	10	58	38	8	22	29
Sn		39				30	(750)	17	12	15	19
Cr			<8		<18	(70)	16	10	33	20	33
Mo			25		36	98		85	147	120	78
Ni		22	<8		3	31	48	62	5	3	23
Co	25	3	15		3	430	30	1400	460	89	
Mn		890	405		390	590	100	60	315	210	370
As			323		489	107	190	110	150	148	220
P						640			420	70	380
V						51			21		36
Carbon		2500[a]					1800	3300	3200	2100	
Spreading Rate (cm/yr)	5.8	5.8	5.8	6.0	6.0	10.0	16.0	16.5	7.0	7.0	
State of Formation	inactive	active	active	inactive	active	active	active	active	inactive	active	
Temp. of Formation	low	low	high?	high	high	high	high	high	high?	high?	

[a] CO_2 not included in mean value;
JFR = Juan de Fuca Ridge; EPR = East Pacific Rise; GRZ = Galapagos Rift Zone

Table 8.2. Compositional variability of a adjacent sulphide deposits (from Scott, 1987)

	Mean and range in chemical composition for eight samples from Explorer Ridge	
Wt %	*Average*	*Range*
Zn	9.0	0–34.3
Cu	8.1	1.0–29.3
Pb	0.1	0.01–0.7
Fe	10.8	4.1–20.1
S	26.8	5.2–43.3
SiO$_2$	19.2	1.3–31.2
ppm		
Au	0.8	0.7–1.5
Ag	112	0–640

parts of mature deposits. Probably to be of any serious resource potential, seafloor polymetallic sulphides should contain enrichments of precious metals.

A three-dimensional study of the interior of a polymetallic sulphide deposit exposed by faulting on the Explorer Ridge and reported by Scott (1987) illustrates the wide compositional range that can be found vertically within adjacent deposits (Table 8.2). Scott considers the range of values in Table 8.2 to be caused mainly by vertical mineralogical zoning of the deposits, and is similar to that seen in many ancient volcanogenic massive sulphide deposits on land. It is evident, therefore, that meaningful resource assessments of seafloor polymetallic sulphide deposits cannot be undertaken until much more information is available on the vertical variation in the composition of these deposits. Sulphide deposits appear either copper-rich or zinc-rich, depending upon which stage of chimney development they are taken from, and thus, as mentioned above, some of the reported geographical differences may simply be an artefact of the sampling.

8.1.3 Tectonic setting of mid-ocean ridge polymetallic sulphide deposits

According to Scott (1987), all seafloor sulphide deposits discovered up until 1987 were in axial or near axial locations on mid-ocean

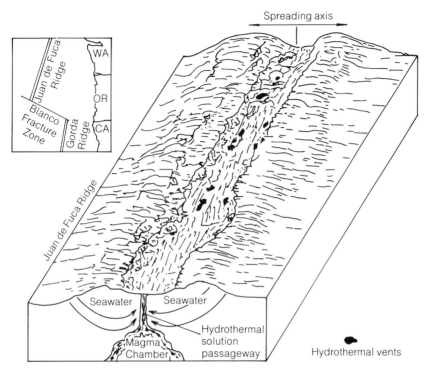

Fig. 8.4 Settings of sulphide deposits on the Juan de Fuca Ridge (modified from McGregor and Lockwood, 1985).

spreading centres, other than a few located on seamounts. Often the sulphide deposits are concentrated along the faulted margins of the axial rift (e.g. Juan de Fuca, Figure 8.4). Ballard and Francheteau (1982) have reported that in at least two locations, on the East Pacific Rise at 13°N and at the axis of the Galapagos Rift near 86°W, regions of shallow sea floor are the main sites of active hydrothermal discharge (Figure 8.5). In order to account for this, a hypothesis was proposed whereby each accretionary segment of a mid-ocean ridge bounded by two transform faults is fed by a single magma plume. Magma was thought to be injected laterally along rift from the central plume forming a magma reservoir. This was considered to reach its fullest development above the plume, where the heat flow is highest. Because of this high heat flow, the oceanic crust becomes raised and thinner which, according to Ballard and Francheteau (1982) leads to the most vigorous hydrothermal activity because of the high energy content in the system at shallow depth. By contrast,

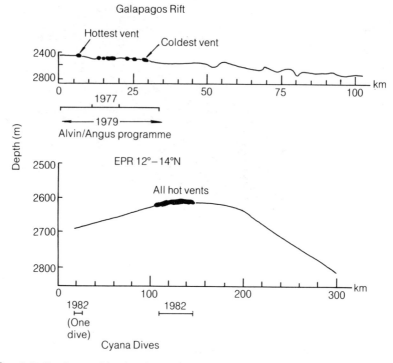

Fig. 8.5 Settings of hydrothermal vents on the Galapagos Rift and on the EPR at 13°N (from Ballard and Francheteau, 1982).

the portions of the rift near fracture zones where the heat flow was thought to be lower were considered to be too cool for the occurrence of major hydrothermal circulation.

The importance of tectonic movements for the initiation and maintenance of ore forming hydrothermal activity has been stressed by Backer and Lange (1987). They report that one of the most common tectonic features of axial rifts are normal faults trending in the rift direction. Further major faulting occurs along the margins of the rift. They note that within a hydrothermal field, most hydrothermal sources are related to these faults which often contain strongly mineralized breccias. In addition, they note that certain sections of spreading centres are characterized by destructive tectonic features accompanied by the formation of piles of talus and collapse pits. Backer and Lange suggest the presence of cycles that start with a constructive magmatic phase and end with the destruction of a raised portion of mid-ocean ridge by faulting and collapse (Figure 8.6).

1. Volcanic phase
Formation of volcanic dome

3. Main rifting phase
Formation of axial grabens

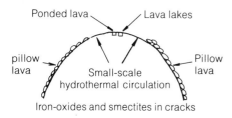

Ponded lava — Lava lakes

pillow lava — Pillow lava

Small-scale hydrothermal circulation

Iron-oxides and smectites in cracks

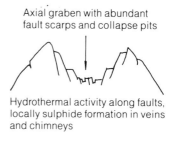

Axial graben with abundant
fault scarps and collapse pits

Hydrothermal activity along faults,
locally sulphide formation in veins
and chimneys

2. Early rifting phase
Fissures and small faults, collapse pits

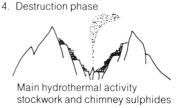

Hydrothermal activity along faults,
oxides, locally some sulphide chimneys

4. Destruction phase

Main hydrothermal activity
stockwork and chimney sulphides

Fig. 8.6 Schematic presentation of phases of activity along the axis of intermediate- to fast-spreading divergent plate boundaries (from Backer and Lange, 1987).

Each phase in this model is thought to have its own specific style of hydrothermalism.

The size of potential hydrothermal ore bodies on the crest of mid-ocean ridges has attracted the attention of several authors. Scott (1987) has pointed out that with the exception of Red Sea deposits, other mid-ocean ridge sulphide deposits are small in size. He reports that among the largest known Pacific deposits are those on the Galapagos Rift at 86°W, on the East Pacific Rise at 13°N and on Explorer Ridge off Canada. However, problems in assessing the thickness of the deposits place considerable constraints on the size estimates given for them, but most appear to be in the range of 3–5 million tonnes. These are about the size of, or slightly larger than, many massive sulphide deposits mined on land.

The work of Backer and Lange (1987) implies that one important constraint on the size of a sulphide deposit on a mid-ocean ridge is the amount of rock that can be leached to provide the metals in the deposit. They note that to generate the 2 million tonne zinc deposit in the Atlantis II Deep of the Red Sea from a basaltic source rock containing 10 ppm zinc, 67 km^3 of rock would be required assuming a 10% transfer efficiency and negligible loss of zinc from the deposit area. They note also that a similar sized ore body on the East Pacific Rise would require a much larger volume of source rock because much of the metal load in the hydrothermal fluids would be dispersed in a hydrothermal plume (Converse *et al.*, 1984; see below) and precipitate in metalliferous sediments widely dispersed around the sulphide deposits. They imply a limitation on the size of the sulphide deposits, based on the size of areas of the ocean crust that can be leached in one convection cell. However, they also point out that studies of land-based ore deposits suggest that some deposits may originate from remobilization of older deposits and thus the energy and metal input of several magmatic/tectonic cycles could build up one major sea floor deposit.

According to Lange (1985a), estimates of the number of hydrothermal fields along mid-ocean ridge spreading centres that are capable of forming massive sulphide deposits range from a maximum of 1 per 3 km to a minimum of 1 per 100 km of ridge crest. However, the size and distribution of the sulphide deposits on mid-ocean ridges may not be entirely related to hydrothermal circulation conditions at the present time. As mentioned, according to Backer and Lange (1987), studies on land-based metallogenic provinces suggest that some deposits or parts of them could originate from remobilization of older deposits.

Converse *et al.* (1984) have noted that the greater proportion of hydrothermal discharge on mid-ocean ridges is lost to the water column as the hot solutions disperse as a hydrothermal plume. Only a small percentage of the total amount of material discharged from a given vent is localized in the immediate vent area. Therefore, according to Lange (1985a), large ore bodies can only develop if hydrothermal solutions are forced to deposit their metals near the site of discharge. He reports favourable settings for this as fault scarps bordering deep axial valley margins at intermediate to fast spreading centres, axial troughs formed during early stages of ocean basin evolution such as occur in the Red Sea, and sediment buried spreading centres.

8.2 SUBMARINE HYDROTHERMAL MINERAL DEPOSITS AT CONVERGENT PLATE MARGINS

Much less is known about hydrothermal mineral deposits at convergent plate margins than on mid-ocean ridges. However, enough is known to demonstrate that the nature of these deposits is

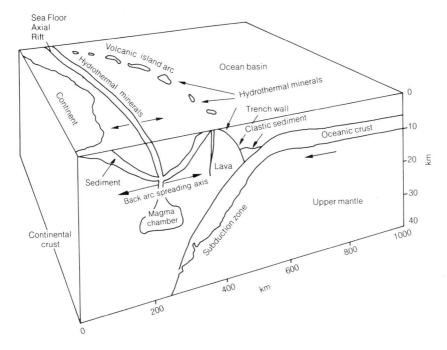

Fig. 8.7 Convergent plate margin with the most common locations of hydrothermal mineral deposits (modified from Cronan, 1980).

similar in its essential features to that on the ridges, namely the occurrence of sulphides, silicates and oxides. Nevertheless, the tectonic situations in which hydrothermal activity can occur at convergent plate margins are more numerous than on mid-ocean ridges and include volcanic island arcs, back-arc basins, seamounts and rifted continental margins. Probably quantitatively the most important of these settings are the back-arc basins (Figure 8.7). Cronan (1980) considered these among the most favourable settings on the ocean floor for the formation of large high grade sulphide deposits.

The origin of hydrothermal mineral deposits at convergent plate margins overall is generally thought to be similar to that at mid-ocean ridges, namely the leaching of sub-surface rocks by circulating hydrothermal solutions, with possibly some magmatic component. However, whereas the composition of mid-ocean ridge rocks is predominantly uniformly basaltic, a much greater variety of rock types occurs at convergent plate margins which can potentially give rise to a much greater variety of hydrothermal solution compositions by rock—water interaction. For example, relatively high amounts of gold and silver and other valuable metals have recently been found in convergent plate margin hydrothermal deposits, as described below. Rare metal enrichments in their sulphides as compared with mid-ocean ridge varieties would be in accord with the more variable composition of the rocks available for leaching than those on mid-ocean ridges. However, different physical and chemical processes at convergent plate margins and on mid-ocean ridges, and between arcs and back-arc basins, can also help to account for compositional differences between hydrothermal deposits in these different settings.

That large hydrothermal mineral deposits can occur on the sea floor at convergent plate margins is demonstrated by the occurrence in the Solomon Islands of massive sulphide overlain by a layer of laminated iron oxide-rich sediments and sinter containing anhydride, barite, opaline silica and pyrite (Taylor, 1974). These deposits are now on land, but the succession appears to represent the zonation from sulphide to oxide deposition on the sea floor which has subsequently been uplifted. Many parts of the western Pacific host hydrothermal deposits of one sort or another, and this area is likely to be one of the most fruitful regions for further exploration for these deposits on the sea floor (see case study below).

Urabe (1989) has recently provided an overview of hydrothermal mineralization in the north-west Pacific, near Japan (Figure 8.8). Barite-rich sulphide chimneys occur on the summits of the ridge seamounts of the Mariana Trough back-arc spreading centre at 18°N. In the Sumisu Rift and Okinawa Trough, hydrothermal activity occurs in rifted back-arc crust. In the Sumisu Rift, the chimneys consist of barite and silica and were found on the flanks of rhyolite lava domes. In the Okinawa Trough, large massive sulphide/barite deposits were found in a submarine depression known as the Izena Cauldron, also studied by Halbach *et al.* (1989) (see below). Small veins of base metal sulphide were found in altered andesite in the wall of a submarine caldera on the Kaikata Seamount, while in one of the frontal volcanoes of the Bonin Arc, anomalies of gold and silver suggested that the veins are analogous to disseminated epithermal gold deposits. These are commonly observed in association with subaerial andesitic volcanoes in the western Pacific. Urabe (1989) correlated these mineral deposits with on land Besshi, Kuroko and epithermal deposits. A summary of the deposits and their tectonic association is given in Table 8.3.

Of particular interest because of their high noble metal content and because they represent the first massive sulphide deposits found in an

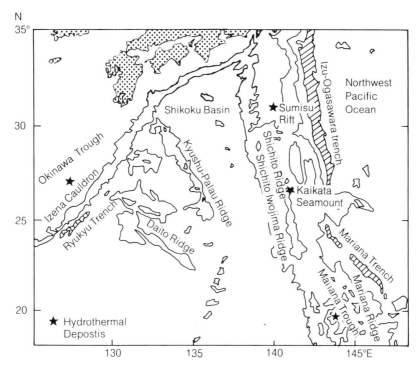

Fig. 8.8 Locations of hydrothermal mineral deposits in the north-west Pacific.

Table 8.3. Tectonic associations of hydrothermal deposits in the N.W. Pacific (after Urabe, 1989)

Location	Tectonic setting	Minerals	Volcanism
Izena Cauldron (Okinawa Trough)	Back-arc rift in continental crust	ZnS, PbS $BaSO_4$ (chimney)	Dacite, rich in volatiles
Sumisu Rift, Bonin Arc	Back-arc rift in island arc crust	$BaSO_4$, SiO_2 (chimney)	Bimodal MORB Rhyolite
Kaikata Caldera Bonin Arc	Volcano on volcanic front	FeS_2, $CuFeS_2$ ZnS (vein)	Two pyroxene andesite
Mariana Trough (Mariana arc)	Back-arc spreading centre	ZnS, PbS $BaSO_4$ (chimney)	MORB

inter-continental back-arc basin are sulphide deposits described by Halbach *et al.* (1989) from the 'Jade' hydrothermal field in the Okinawa Trough. This is a back-arc basin forming by extension within continental lithosphere. The Jade hydrothermal field occurs in the Izena Cauldron at 27°15.0'N, 127°04.5'E. Four different types of sulphide deposits were recovered, identified as two varieties of massive sulphide, the first black-grey and the second olive-green to grey, a stockwork mineralization and a sulphide bearing sediment layer. The massive sulphides were rich in zinc and lead, up to 40% and 19.5% respectively, but contained less than 1% copper. Their concentrations of gold and silver, 0.65% silver and 9.8 ppm gold, are among the highest values of these metals reported to date in seafloor sulphide deposits. Halbach *et al.* (1989) assumed that remobilization of gold from earlier formed higher temperature sulphides had taken place, followed by its concentration in the more oxidized outer portions of the deposit, and concluded that the 'Jade' hydrothermal deposits are analogous to Kuroko-type massive sulphides. Further details on the 'Jade' hydrothermal deposits have been given by Halbach (1989). The stockwork mineralization was found to contain sphalerite, silver and antimony-bearing galena, barite, pyrite and chalcopyrite. The sulphide-bearing sediments occur at about 70–80 cm depth in cores adjacent to the 'Jade' hydrothermal field and contain sphalerite, pyrite and barite with up to 1.13% silver.

8.3 SUBMARINE HYDROTHERMAL MINERAL DEPOSITS ON SEAMOUNTS

According to Scott (1987), seamounts, often but not always those with calderas, are an important location for seafloor sulphides. The reason given is the abundant deep fissures within calderas which facilitate the circulation and flow of the hydrothermal fluids necessary for sulphide formation. According to Backer and Lange (1987), such volcanoes are likely to be the result of an excess of magma production, leading to prolonged high heat flow and therefore extended hydrothermal activity. Many seamount hydro-thermal deposits are on seamounts close to mid-ocean ridges and may well have been initiated at the ridge axis. In a sense, these can be classified as mid-ocean ridge hydrothermal deposits. However, others are associated with oceanic hotspots or other features and important examples of these include the Palinuro seamount deposits in the Tyrrhenian Sea (Minetti and Bonavia, 1984), the Loihi

Seamount deposits off Hawaii, and seamount deposits in the south-eastern Pacific.

Volcanogenic massive sulphides, mainly copper-rich and mixed with metal bearing sediments, were reported from a series of craters at around 1000 m depth, on the Palinuro Seamount, Tyrrhenian Sea (Figure 8.1) by Minetti and Bonavia (1984). The minerals consist of indurated sulphide crusts containing the copper minerals luzonite, tennantite-tetrahedrite and native copper, associated with pyrite and barite. Native silver, bismuthinite and stibnite were also found.

Loihi Seamount, off Hawaii, is at the southernmost extent of the Hawaiian hotspot (Figure 7.5) and is an active young submarine volcano that is probably the site of an emerging Hawaiian island (Malahoff, 1982). Dredging of the rim and summit crater of the seamount at about 1000 m has recovered several types of hydrothermal deposits. According to De Carlo *et al.* (1983), these suggest a precipitation sequence in which iron silicates ranging in composition from iron-montmorillonite to nontronite have precipitated along with iron oxide, to be followed by iron oxide and silica deposition. Formation temperatures were thought to be in the range of 31°C to 57°C for the iron silicates. No polymetallic sulphides were recovered, but these were suspected as possibly occurring beneath the surface hydrothermal deposits within the body of the seamount. The surface deposits would be the equivalent of the middle of the range in the hydrothermal fractionation sequence of sulphides, silicates, and oxides described on p. 137.

Additional mid-plate hotspot hydrothermal occurrences have been recorded in the south-eastern Pacific, in the vicinity of Pitcairn Island, Teahitia and on the Macdonald Seamount (Figure 4.1) (Stoffers, 1987). The fluids discharged are similar to those on Loihi Seamount. The hydrothermal minerals are dominated by iron oxyhydroxide crusts of restricted areal extent, and are low in transition metals other than iron. They are thought to be low temperature hydrothermal deposits. Submersible operations using CYANA demonstrated the presence of iron-rich 'nontronitic' chimney fields on Teahitia Seamount which formed at low (<40°C temperatures. However, the possibility exists of sulphide mineralization on or in Macdonald Seamount, as a sea surface slick sampled during an active submarine eruption was found to contain particulate sulphide phases.

Interestingly, with the exception of Palinuro Seamount, the other non-mid-ocean ridge seamounts sampled so far for hydrothermal deposits appear to host on their surfaces only silicates and oxides.

Sulphides are rare or absent. Depth related factors such as pressure are unlikely to account for this as the depths of the summits of Palinuro Seamount (sulphide bearing) and Loihi Seamount (non-sulphide bearing) are very similar at around 1000 m. Rather, the isolation of the seamounts, surrounded as they are by seawater, coupled with their often fractured structure, probably permit the ingress of substantial amounts of seawater which cool the hydro-thermal solutions sub-surface leading to the precipitation of sulphides within the seamounts. Only low temperature hydrothermal solutions would generally discharge at the seamount summits unless sub-surface cooling is prevented by some mechanism as evidently happens in Palinuro Seamount. Similar constraints probably apply to submarine volcanoes in island arcs (Cronan *et al.*, 1982; Moorby *et al.*, 1984; Cronan, 1989b) and may even apply on mid-ocean ridges if the seamounts are sufficiently isolated and elevated.

8.4 CASE HISTORIES

Three case studies illustrating submarine hydrothermal deposits in diverse tectonic settings have been chosen to illustrate the variability of these deposits.

8.4.1 Hydrothermal deposits in the Red Sea

Although formed at a spreading centre, hydrothermal deposits in the Red Sea are sufficiently distinctive from other spreading centre deposits described to date to warrant a separate consideration. Unlike mature mid-ocean ridge deposits, Red Sea deposits are formed on an embryo spreading centre with continental margins close on either side of it (Figure 8.1). This results in the hydrothermal deposits being ponded deeper than the general level of the surrounding sea floor, in contrast to mid-ocean ridges where the deposits are often on the most elevated parts of the sea floor (Ballard and Francheteau, 1982). Their setting also admits of the possibility of seawater circulation through continental as well as oceanic rocks prior to their discharge on the sea floor as hydrothermal solutions and thus they are able to pick up a wider variety of constituents than hydrothermal solutions which just circulate through oceanic crust. The high salt content of the solutions is thought to be a result of this phenomenon, namely the circulation of the hydrothermal solutions through evaporite beds. Additionally, some of the high metal concentrations in Red Sea deposits are thought to result from hydrothermal solutions circulating through metalliferous black shales (Manheim, 1974).

It was during the Swedish Deep Sea Expedition in the late 1940s that the first indications of hydrothermal activity in the Red Sea were found. These were confirmed in the mid 1960s when brine-filled depressions were detected in the central Red Sea. Furthermore detailed investigations showed the presence of a brine in the Atlantis II Deep (Figure 8.2) with a temperature of 56°C, overlying sediments rich in iron, manganese, zinc, copper, cadmium, lead and silver (Miller *et al.*, 1966). The metal-rich sediments in the Red Sea, often together with overlying hydrothermally enriched brines occupy the central graben of the Red Sea rift and appear to be preferentially concentrated where this is offset by faults or fractures (Bignell, 1975) (Figure 8.9). Bignell *et al.* (1976b) divided the Red Sea hydrothermal metalliferous sediments into a number of different facies characterized by different mineral phases. These are the oxide facies, sulphide facies, sulphate facies, carbonate facies, and silicate facies, and their distribution in the different Red Sea deeps were described (see Cronan, 1980 for a summary).

According to Girdler and Styles (1974), the Red Sea has evolved through at least two stages of sea floor spreading, the first of which was 41–34 million years ago and the second began about 4–5 million years ago and is still continuing. During the intervening inactive period, evaporites were deposited (Gass, 1977). As mentioned, it is the leaching of these evaporites that has given rise to the high salinity brine that characterizes many of the deeps within the Red Sea. This increase in salinity of the hydrothermal solutions enhances their capacity to transport metals as chloride complexes and also provides a density stratification within the brine pools after discharge which serves to pond the hydrothermal precipitates and prevent their dispersion away from the sites of hydrothermal discharge, as generally occurs in hydrothermal plumes on open ocean mid-ocean ridge crests. Thus the brines play a very important role in both forming the hydrothermal deposits within the Red Sea and also in concentrating them.

The brines and hydrothermal sediments in the Red Sea which have received the greatest attention are those in the Atlantis II Deep, because it is there that the deposits exhibit the greatest economic potential. Published data on Atlantis II Deep brines have been summarized by Lange (1985b) and more recently updated by Hartman (1985). Exit temperatures for the brines are thought to be in the range of 230°C to 400°C, similar to those recently measured in hydrothermal vents on the East Pacific Rise. According to Lange

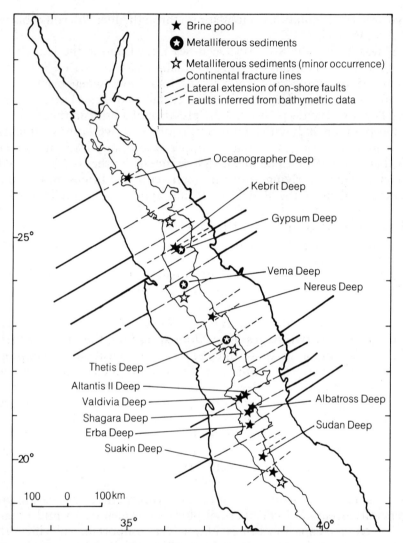

Fig. 8.9 Locations of major hydrothermal occurrences in the Red Sea and their relationship to faulting (from Bignell, 1975).

(1985b), the Atlantis II Deep brines are enriched relative to Pacific hot springs in sodium, chlorine, potassium and calcium, most likely as a result of the leaching of the evaporites referred to above. Higher concentrations of manganese in the brines than in open ocean mid-ocean ridge hot springs are explained by the dissolution of manganese oxides settling down from the more oxidizing waters above the brine (Figure 8.2). Hartman (1985) has recently drawn attention again to the stratified nature of the brine in the Atlantis II Deep, which contains a lower brine and an upper brine and a transition zone between brine and Red Sea waters. Chemical reactions taking place in the brine are summarized in Figure 8.2.

The sequence of metal precipitation in the Atlantis II Deep is as follows. Iron, copper and zinc start to precipitate first as sulphides from the lowermost brine in the vicinity of the discharge vents. This is a result of temperature decrease between vent exit temperatures and the ambient temperature of the lowermost brine which is in the vicinity of 60°C. Chalcopyrite precipitates first, next sphalerite and finally iron sulphide (Lange, 1985b). Dissolved silica reacts with iron oxide settling down from above to give iron silicate phases (Cole, 1983). The iron oxides precipitate mainly in the lower portion of the transition zone between the brines and normal Red Sea water as a result of oxidation of Fe^{2+} to Fe^{3+} (Lange, 1985b). Finally, manganese oxides and oxyhydroxides form at the higher Eh and pH found in the upper portion of the transition layer and overlying Red Sea waters. Much of this last group of precipitates is carried away from the immediate vicinity of the Deep by bottom currents and is sedimented around the margins of the Deep not overlain by brine. Bignell *et al.* (1976a) have reported a zone of sediments enriched in manganese oxides up to 11 km away from the Deep formed by this process.

Hydrothermal sedimentation within the Atlantis II Deep has been taking place for at least 25 000 years (Backer and Richter, 1973). This has led to the deposition of various sediment facies at different times (Figure 8.10). The lowermost unit within the Atlantis II Deep is the detrital oxidic pyritic zone (DOP) which was deposited directly above basement and consists primarily of detrital carbonates, clays and silicates, with minor contributions of limonite and pyrite. Above this is the lower sulphide zone (SU_1) which represents the first period of stable widespread hydrothermal activity within the Deep. It contains sulphides, consisting primarily of sphalerite, pyrite, and chalcopyrite, together with other minerals such as iron silicates, manganosiderite and anhydrite (Lange, 1985b). Above this is the central oxidic zone (CO) which is characterized by oxidized material and contains goethite, haematite and silicates as its most abundant minerals. Above this is the upper sulphidic zone (SU_2) which represents a return to reducing conditions in the Deep and contains mineralogical assemblages similar to those in the lower sulphidic zone. Finally, the uppermost unit in the Atlantis II Deep is the amorphous silicate zone (AM) which is being formed at the present

m	Unit	Mineral content	Age [y]	Solid Matter [%]	Zn	Cu	Ag	Fe
					[%]in solid matter		[ppm]	[%]
0 —	**AM**	Amorphous Fe–silicates + Sulphides		7	3.1	0.5	62	33
3 —	**SU₂**	Sulphides + Fe–silicates		10	3.1	0.5	56	27
5 —	**CO**	Limonite + haematite + manganite + Fe–silicates	5000	15	1.2	0.4	29	31
8,5 — / 10 —	**SU₁**	Sulphides + silicates	Holocene 10 000	22	3.1	0.8	58	24
	DOP	Limonite + marl + pyrite	Pleistocene	36	0.7	0.2	19	22
15 —	**BASEMENT**	Basalt						

Fig. 8.10 Stratigraphy, mineral content and chemical variability with depth in sediments from the Atlantis II Deep (from Lange, 1985b).

day and consists mainly of poorly crystalline iron silicates together with sulphides and anhydrite. These sediments are likely to be the first hydrothermal mineral deposits to be mined on the sea floor. Their resource potential will be considered on p. 173.

8.4.2 Hydrothermal sulphide deposits on the East Pacific Rise near 13°N

The East Pacific Rise near 13°N has been the site of numerous detailed studies on submarine hydrothermal mineral deposits since 1981. More than one hundred submersible dives have been conducted in an area bounded by 12°40′N, 103°56′W and 12°52′N and 104°56′W (Figure 8.11). Hydrothermal deposits have been obtained from four different tectonic settings in the area, active smoker vents on the bottom of the axial rift, inactive and faulted areas on the top of the walls of the graben flanking the rift, active vents on an off-axis structure which is a marginal 'high' developing into a seamount, and lastly on a seamount located 6 km east of the axis. Nine types of hydrothermal deposits have been described, a greater variety than in any comparable submarine hydrothermal area other than the Red Sea. Although this area is not part of an Exclusive Economic Zone,

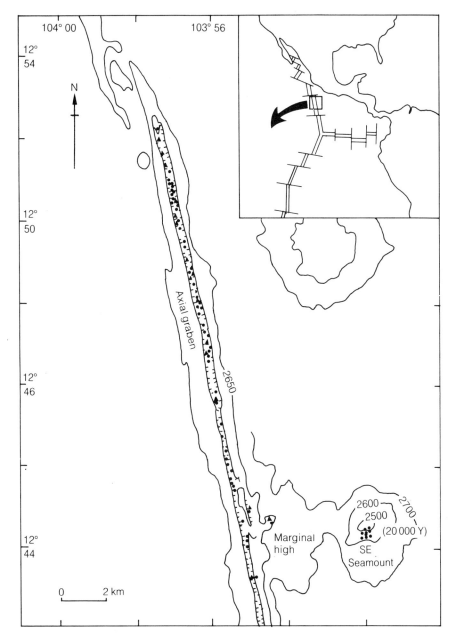

Fig. 8.11 Locations of hydrothermal deposits on the East Pacific Rise near 13°N (from Fouquet *et al.*, 1988): ● = Inactive vents ▲ = Active vents.

the great variety of hydrothermal deposits that it displays encompasses most if not all of the types found in submarine hydrothermal areas in Exclusive Economic Zones off western North America. In none of those areas, however, has a greater variety of deposits been reported as in the 13°N area, and therefore its inclusion in the present work is felt to be justified. A comprehensive report on the hydrothermal activity and mineral deposits in the 13°N area has been presented by Fouquet *et al.* (1988) which is summarized below.

The axial zone of the East Pacific Rise in the 13°N area contains an axial graben between 200 and 600 m wide, 20 to 50 m deep, with an average water depth on its floor of 2630 m. The middle part of its flat floor contains a fissure 400 m long and several metres wide which acts as a focus for hydrothermal activity. On its western side, there are a series of small horsts and grabens parallel to the axial graben, while on its eastern flank, there are two seamounts. The second of these, the so-called South Eastern Seamount, is located 6 km from the axis and is connected to it by a marginal high which is interpreted as a young seamount whose growth has continued since it formed at the axis (Figure 8.11).

The hydrothermal deposits in the axial graben are thought to be related to recent fissuring and post-date the youngest lava flows. Recovered samples consist of young immature sulphides characterized principally by copper and zinc chimneys, and porous iron and/or zinc-rich sulphides growing directly on the basalt substrate. In this area, chimneys are very active with heights of up to 25 m, and diameters of up to 3 m. Most of the zinc-rich chimneys show evidence of late copper sulphide precipitation suggesting an increase in temperature during chimney growth. Observations were held to indicate that along the central tectonically active zone, the hydrothermal activity is increasing in intensity.

Hydrothermal deposits on the walls of the graben comprise the second line of hydrothermal activity, 150 m east of the axial line. However, the deposits are inactive, faulted and generally located on the top of the graben wall, and are thought to correspond to a past hydrothermal episode during an earlier stage of graben formation. However, small chimneys at the foot of one of the fault scarps were observed, demonstrating continuing weak hydrothermal activity along the wall after graben formation. Mineralogical studies on the graben wall deposits show that in contrast to those from the active axial line, the latest mineral formation there was associated with decreasing temperatures. Crusts mainly composed of opaline silica with minor barite were formed during the waning stage of the hydrothermal activity. Mounds are composed of sulphides inter-mediate between porous and massive types. The deposits are more mature than those at the axis, chimney spires are scarce and the mounds are more fully developed.

The young seamount known as the 'marginal high' is the only off-axis area where active hydrothermalism is taking place at 13°N. It exhibits two main deposits: one at its top where the deposit is one of the largest in the 13°N area, and one on its western side associated with a fault at the eastern limit of the graben. The deposits are iron and copper-rich sulphides which are overlain by porous immature sulphides and active chimneys. Lead isotope dating by Lalou *et al.* (1985) has indicated an age of about 2000 years for the deposits. Fouquet *et al.* (1988) point out that as hydrothermal activity is still occurring on the marginal high, it can be concluded that it takes at least 2000 years to form a massive sulphide deposit. This is similar to the time estimated for the formation of comparable sulphide bodies mined on land.

The South Eastern Seamount at 13°N is located 6 km from the axis, has a basal diameter of about 6 km and is about 350 m high (Figure 8.11). The hydrothermal deposit on this seamount is the biggest in the 13°N area, and indeed one of the biggest so far recorded on the whole of the East Pacific Rise, 3.8 million tonnes (Hekinian and Bideau, 1985). Electrical resistivity measurements (Francis, 1985) have indicated that the thickness of the deposit can reach at least 10 m, although it is only about 1 m thick in places. Fouquet *et al.* (1988) consider the massive sulphide body to be a lens rooted by stockwork on a large fault, and this fault is considered to be the pathway for the hydrothermal discharge. Near the fault, the massive sulphides are copper-rich while away from it the thickness of the body decreases and the samples become more iron-rich, with the sulphides being oxidized in contact with seawater and forming gossans. Sampling and submersible observations on the South East-ern Seamount show several different types of hydrothermal deposits. These include copper-rich sulphides which are preferentially located along the fault, iron-rich sulphides which are observed at many outcrops, opal and quartz associated with iron oxides near the top of the seamount, manganese crusts associated with nontronite also on the top of the seamount, and widely distributed oxyhydroxides representing oxidized sulphide materials. Like on the marginal high, zinc-rich samples are rare or absent. In all, the South Eastern Sea-mount deposits represent the bulk of the hydrothermal deposits found in the 13°N area and are considered by Fouquet *et al.* to be more mature massive sulphides than those found on the axis.

Fouquet *et al.* (1988) report that mineral assemblages and deposit compositions vary considerably throughout the 13°N area. Ranges are given in Table 8.4. Iron hydroxides which result from the oxidation of iron sulphides are enriched in nickel and depleted in cobalt compared with the sulphides. However, lead, arsenic, silver and cadmium occur in these deposits in concentrations similar to those in non-oxidized iron sulphides, and one important feature of them is the enrichment in gold from 2.9 to 4.1 ppm which is concentrated during weathering of the sulphides in contact with seawater. Hydrothermal manganese crusts contain low copper, cobalt and nickel and high manganese/iron ratios, characteristic of these sorts of deposits elsewhere in the oceans. Other trace elements occur in them in small amounts compared to those found in sulphides. However, gold is present in them in concentrations of 1.2 ppm.

Statistical analysis of the geochemical data on the 13°N deposits led Fouquet *et al.* (1988) to propose a fractionation of three different groups of elements from the hydrothermal fluid on precipitation. The first group, copper, iron, selenium, cobalt, calcium and strontium, was thought to be characteristic of a high temperature mineral forming association and to occur both in axial and off-axial hydrothermal deposits. The second group, zinc, cadmium, silver, arsenic and lead, occurs in zinc-rich chimneys and iron-zinc porous sulphide mineral deposits and is located only in the axial graben, and is thought to be characteristic of a medium temperature precipitation. The third group, silicon and barium, forms opal and barite as a late stage low temperature precipitate.

Of the potentially economically valuable minor elements, gold has been found to be enriched up to 4.1 ppm in iron hydroxides and in manganese oxide phases, compared to its mean value in the sulphide mineral assemblages of close to 0.5 ppm (range 0 to 3.0 ppm). These values are thought to result from a high gold content of the primary hydrothermal fluid. Similar values have been obtained in the deposits by Japanese workers (Konagi, personal communication, 1989).

Silver occurs in two settings. First it is held in dendritic sphalerite in the outer part of zinc-rich chimneys, and can be present there in amounts of 0.2%. Second, in the cores of the chimneys, the silver is held equally in zinc sulphides, chalcopyrite, intermediate solid solution (ISS) and pyrrhotite, but in lesser concentrations of only about 0.05 to 0.09%. Platinum group elements have been sought in the deposits, but values were lower than the detection limit which was 5 ppb.

Fouquet *et al.* (1988) have attempted to outline the physicochemical conditions of sulphide formation in the 13°N area on the

Table 8.4. Compositional variability in hydrothermal deposits from the EPR at 13°N in weight % (after Fouquet et al., 1988)

	Zn	Cu	Fe	Ca	S	Mn	Pb
Cu chimney	0.02– 2.95	0.67–32.2	2.65–31.9	0.01–25.44	23.5–34.5	0.01–0.055	0 –0.022
Zn chimney	2.6 –46.9	0.1 – 2.92	10.6 –39.75	0.04–19.1	31.7–45.8	0.01–0.055	0.027 –0.28
Porous sulphide	0.03–33.8	0.70–20.4	20.0 –43.8	0.05– 1.90	36.5–48.3	0.01	0.018 –0.084
Massive sulphide	0.03– 0.75	0.22–28.8	33.8 –47.0	0.01– 0.20	34.3–51.0	0.01	0.013 –0.025
Fe oxides	–	–	24.5 –30.09	2.14– 3.15	–	–	0.0013–0.0057
Mn oxides	0.03	–	1.13	1.21	0.10	45.1	0.0001

basis of the mineralogical and geochemical data reported. The nature of the initial conditions of precipitation can be estimated by observation of mineral assemblages occurring in plume particles which result from rapid quenching after discharge. These have been found to be rich in pyrrhotite, chalcopyrite, isocubanite and anhydrite. On decreasing temperature, the pyrrhotite is destabilized and replaced by an intermediate product (Py-Ma) and sulphate. The temperature decrease results from a mixing of the hydrothermal end member fluid with seawater. In these conditions, hydrothermal fluids holding at least 1 ppm zinc and less than 1 ppm copper are over-saturated as far as the minerals sphalerite and chalcopyrite are concerned. Insulation and/or the rewarming of the hydrothermal fluid relative to seawater leads to an increase in zinc sulphide solubility as zinc chloride ($ZnCl_4^{2-}$) and subsequent zinc mobilization. This would account for the preponderance of copper-iron-rich mineral assemblages in the samples collected in the 13°N area, and could account for the apparent absence of zinc from the older massive sulphide deposits off axis.

8.4.3 Hydrothermal mineral deposits in the south-western Pacific

Unlike in the case of the Red Sea or on the East Pacific Rise, high grade hydrothermal mineral deposits have not been found in abundance in the island arcs and back-arc basins of the south-western Pacific. However, the first submersible operations in the region during 1989 did locate sulphide deposits in areas where they had been expected on the basis of the areas being known to contain other hydrothermal minerals such as iron silicates and iron and manganese oxides normally associated with sulphide. Based on the widespread distribution of these other hydrothermal minerals throughout the south-western Pacific it can be deduced that high grade polymetallic sulphides are likely to be abundant there.

Initial scattered reports of submarine hydrothermal activity in the south-western Pacific were consolidated into a review of hydrothermal activity in the area by Cronan (1983), which indicated that hydrothermal minerals could be expected in two main tectonic settings, the volcanic arcs and the back-arc basins. Likely deposition areas included the plate boundary in the Bismarck Sea, the spreading centre in the Woodlark Basin, submarine volcanoes off Vanuatu, the Tonga-Kermadec Ridge, and the spreading centres in the Lau and North Fiji Basins. Subsequent work on south-western Pacific hydrothermal mineral deposits has largely been concentrated in these areas.

The Bismarck Sea (Figure 8.12) is divided by the Manus–Willumez Rise into the New Guinea Basin to the west and the Manus basin to the east. According to Taylor (1979), this area is a back-arc basin with respect to the New Britain-Arc Trench system. Cronan (1983) and Coward and Cronan (1985) evaluated the possibility of hydrothermal mineral deposition in the Manus Basin on the basis of geochemical anomalies in sediments and proposed hydrothermal activity close to the transform and spreading centres (Figure 8.12). More recent work in the Manus Basin has shown that inactive chimney vents similar to those on the East Pacific Rise occur there in association with methane anomalies (Both *et al.*, 1986; Craig and Poreda, 1987) (Figure 8.12). In addition, hydrothermal manganese crust deposits have also been discovered well away from the site of the supposed spreading centre, as well as near to it (Bolton *et al.*, 1988). On the basis of evidence of submarine hydrothermal activity collected to date in the Manus Basin, the most likely settings for high grade hydrothermal deposits such as polymetallic sulphides are to be found in the axial region of the spreading centres in the central and eastern Manus Basin. More limited possibilities may exist in the transforms linking these features.

The Woodlark Basin is thought to be a zone of active seafloor spreading lying in the southern Solomon Sea between the south-facing Bismarck Volcanic Arc and the Deep Coral Sea Basin but is not a back-arc basin *sensu-strictu* (Binns *et al.*, 1987; 1990). It is spreading at a half rate of as much as 3.5 cm per year and has propagated westwards at about 12 cm per year into a region of continental crust (Taylor, 1984). In its eastern part there is a hydro-thermal methane anomaly at Scripps Papatua Station VI-6 (Craig and Poreda, 1987) (Figure 8.13). Binns *et al.* (1987) considered the western part of the Woodlark Basin to be a possible modern analogue of the settings in which ancient volcanogenic massive sulphide ore bodies were formed. Hydrothermal deposits have been dredged on the Franklin Seamount in the PACLARK area (a joint Australian-Canadian hydrothermal study area) (Figure 8.13) con-sisting of hundreds of centimetre size fragments of indurated orange to red-brown deposits composed of a silica bearing iron oxyhydrate with an accessory manganese phase and micron size barite grains (Binns *et al.* 1987). These authors concluded that venting of hydrothermal fluids, probably at relatively low temperature, was the essential factor in the formation of this deposit, and considered it to be the equivalent of some of the iron silica deposits surrounding

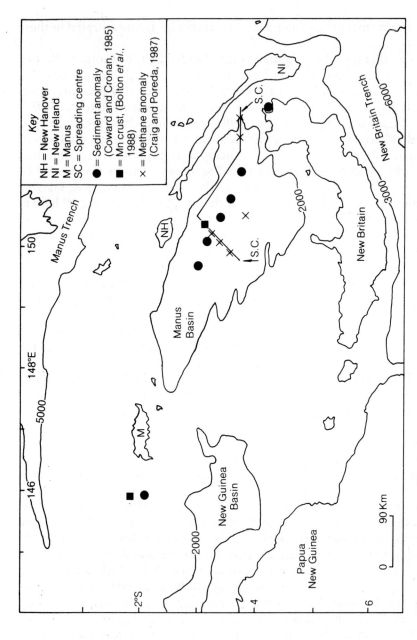

Fig. 8.12 Locations of hydrothermal activity in the Bismarck Sea (modified from Cronan, 1989b).

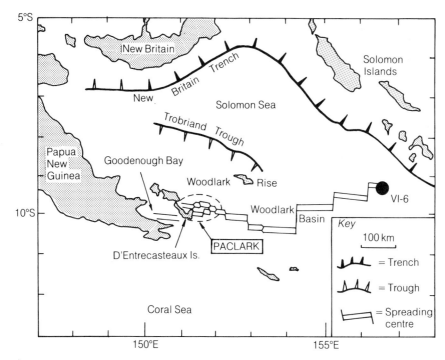

Fig. 8.13 The Woodlark Basin showing the locations of the hydrothermally active PACLARK area and a methane anomaly at Station VI-6.

sulphide mounds and high temperature vent fields on mid-ocean ridges. Subsequently, Binns *et al.* (1989, 1990) reported enrichments of arsenic, antimony, silver, copper and zinc in some iron minerals recovered from the PACLARK area. Binns *et al.* (1990) have also reported on dives in the PACLARK area by a Russian submersible which provided evidence of the presence of low temperature Fe-Mn-Si oxide deposits and extinct barite-silica chimneys rich in lead. In the same general region as the PACLARK area, extensional rifting is thought to occur in the Goodenough Bay sedimented basin to the west of Normanby Island (Figure 8.13) and therefore this area must be considered to be an additional hydrothermal target, not only for surface hydrothermal activity, but also for buried and therefore possibly concentrated hydrothermal deposits of the Besshi-type.

The North Fiji Basin is a back-arc basin averaging about 3000 m deep (Figure 8.14). It stands between the New Hebrides subduction zone to the west and the Fiji Platform to the east. It is thought to have originated by oceanic spreading about 8 to 10 million years ago

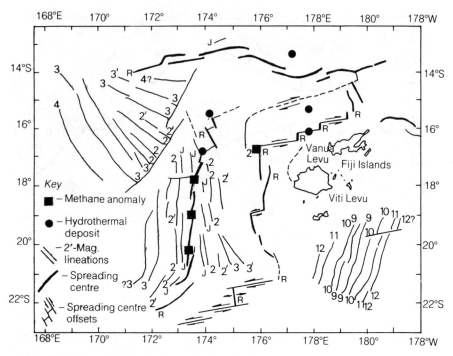

Fig. 8.14 Locations of submarine hydrothermal activity in the North Fiji Basin. (Magnetic lineations and structural data by A. Malahoff).

and has undergone several modifications in spreading activity since that time (Auzende *et al.*, 1988).

According to Auzende *et al.* (1988), the main spreading centre in the North Fiji Basin is located between 173°E and 174°E and is defined by a central magnetic anomaly and lack of sediment cover. This feature has been the main, but not sole (Cronan, 1983), target for hydrothermal mineral exploration in the North Fiji Basin.

Within the axial rift of the North Fiji Basin, between 16°56′S and 17°S and 173°53′E and 173°57′E are some hydrothermal methane anomalies (Figure 8.14) and many yellowish-brown deposits in the deeper parts of the graben. Some irregularly shaped yellowish-red mounds are also present. These mounds are thought to be hydrothermal mineral deposits (Kaiyo unpublished Cruise Report, 1987). More recently, video evidence of a small but active white smoker at 16°59′S, 173°54′E has been obtained (Kaiyo unpublished Cruise Report, 1988). Earlier, during the Sonne 35 Cruise in 1984, hydrothermal sulphide minerals were found in the northern North Fiji Basin (Figure 8.14). They occurred as minute pyrite crystals along

fractures in moderately fresh pillow basalts. In addition, hydrothermal manganese crusts were recovered from localities near the
Braemar Ridge (Figure 8.14). Sediment geochemical studies have
indicated several additional possible locations of hydrothermal activity in the North Fiji Basin (Cronan, 1989b).

The first submersible operations in the North Fiji Basin took place
in mid-1989 and resulted in a considerable extension of our knowledge of hydrothermal deposits there (Auzende *et al.*, 1989). The main
dive site was located in the axial graben at the northern tip of a
spreading axis between 16°58′ and 17°S at 174°E. According to
Auzende *et al.* (1989), extinct hydrothermal sites were found all along
the graben, containing fossil sulphide chimneys, ferromanganese
oxide staining on the rocks and dead shells. In the centre of an
inactive hydrothermal area at 16°59′S, an active vent was discovered
on a 7 m high sulphide mound, capped by a 3 m high accumulation
of anhydrite and named the 'White Lady'. An accumulation of polymetallic sulphide discovered around 16°58′S is among the largest yet
seen on the ocean floor. It is more than 1 km in diameter and where
cut by faulting is up to 40 m thick (Auzende *et al.*, 1989). Its
occurrence bears out previous predictions that convergent plate
margin settings could host large polymetallic sulphide deposits.

The Lau Basin (Figure 8.15) separates the Tonga Ridge from the
Lau Ridge and is an active back-arc basin. Volpe *et al.* (1986) have
proposed that oceanic crust in the Lau Basin has been formed by
spreading on short ridge segments and at numerous seamounts over
the last 3 million years, and that the basin contains propagating rifts,
abandoned rifts, jumped spreading centres and active seamounts. All
these are potential sites for hydrothermal deposits.

Hydrothermal deposits have been reported from a number of
places in the Lau Basin. The first was a barite-opal deposit from the
Peggy Ridge described by Bertine and Keene (1975) (Figure 8.15).
Subsequently, von Stackelberg *et al.* (1985) identified two regions in
the Lau Basin where there is evidence of hydrothermal activity. The
first is in the northern Lau Basin where underwater photography
indicated the presence of manganese encrustations together with
what appeared to be hydrothermal nontronite. The second area was
further south and includes the Valu Fa Ridge where porous hydrothermal nontronite was recovered, together with manganese crusts.
Sulphide mineral impregnations of rocks recovered to the east of
the active spreading zone, together with manganese crusts and
nontronite were also recovered. Subsequent work by von Stackelberg
et al. (1988) located many more hydrothermal deposits in the Lau
Basin (Figure 8.15). In the southern part of the northern Valu Fa

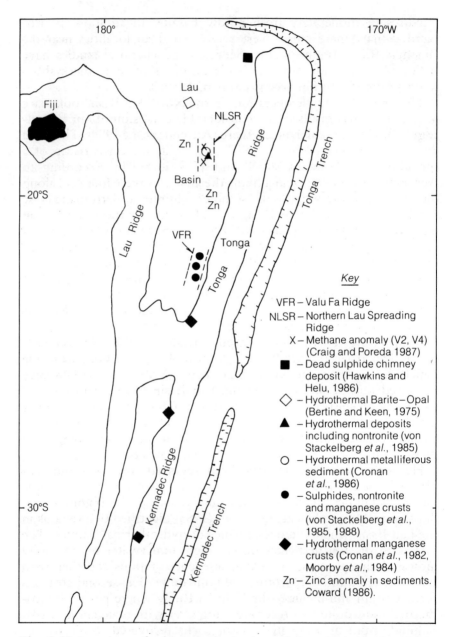

Fig. 8.15 Locations of submarine hydrothermal deposits in the Lau Basin and on the Tonga Kermadec Ridge.

Ridge, they observed nontronite and manganese crusts, and pyrite impregnated rocks on a small seamount 7 km to the east of the ridge. Old sulphide chimneys with attached benthic organisms were seen during deep-tow video/photo sledge operations. One dredge recovered a few lumps of massive sulphide consisting mainly of sphalerite, pyrite, chalcopyrite and galena with associated opal. On the southern Valu Fa Ridge, they observed a large hydrothermal field with a maximum width of about 4 km containing nontronite and manganese oxide crusts.

More recently, a Franco/German research programme in 1989 using the submersible *Nautile* found numerous polymetallic sulphide deposits on the Valu Fa Ridge between 21°25′S and 22°40′S, containing up to 29% copper, 53% zinc, 24% barium and 2.5% lead. Additionally, the submersible discovered the existence of active hydrothermal vents in many places; the highest temperature recorded was 400°C in a vent field showing numerous black and white smokers. Three types of hydrothermal deposits were recovered during the diving operation (Fouquet *et al.*, 1989). First, low temperature deposits consisting of manganese and iron oxides were found to be related to discharge through highly vesicular andesite. Second, medium to high temperature barite/sulphide deposits were observed at many places and were held to represent a hitherto unknown type of hydrothermal deposit on the ocean floor. The most important field was a few hundred metres in diameter consisting of barite chimneys and massive barite boulders mixed with massive sulphides. Third, very high temperature (up to 400°C) black and white smokers were found to be discharging from one of the most hydrothermally active areas yet discovered on the ocean floor. A complete cross-section was sampled through a massive sulphide deposit, including the stockwork. This was the first submersible operation in the Lau Basin designed to search for hydrothermal mineral deposits, and future submersible operations can be expected to find many more such deposits.

In the northern Lau Basin, Hawkins and Helu (1986) have reported an occurrence of hydrothermal sulphides consisting of fragments of a dead black smoker chimney at 15°23′S, 174°41′W (Figure 8.15). The deposits consist of concentric layers of sphalerite, pyrite and chalcopyrite with sphalerite being the main mineral. Subsequent work in the northern Lau Basin by Russian submersibles in early 1990 resulted in the location of large fields of sulphide chimneys (Malahoff, personal communication, 1990).

Unlike the geological settings described so far in this section, the Tonga-Kermadec Ridge is a volcanically active island-arc. However, it has yielded hydrothermal manganese oxide deposits (Cronan *et al.*, 1982; Moorby *et al.*, 1984). The manganese oxide crusts were recovered from three localities (Figure 8.15) and consist of layered to massive oxides composed largely of birnessite but with subsidiary todorokite. Compositionally, the deposits are rich in manganese, up to about 50%, and low in most other metals. However, they do not have the same economic potential as manganese nodules, because of their low minor element content.

Interestingly, the Tonga-Kermadec Ridge manganese crusts appear not to be associated with other hydrothermal deposits. No sulphides or nontronite have been found in their vicinity as often occurs in the vicinity of manganese deposits in back-arc basins, and their associated sediments are normal deep sea oozes without manganese enrichments. There are at least two possible explanations for this. Either, the hydrothermal solutions were extensively cooled sub-surface, resulting in sulphide and silicate precipitation below the sea floor as proposed in the case of some seamount deposits (p. 154) and leaving only low temperature manganese-rich solutions to discharge on to the sea floor, or only low temperature leaching of the volcanic rocks took place resulting in the mobilization of only manganese. In either case, such processes would not be expected to result in massive sulphide deposition on the sea floor. Additional hydrothermal manganese deposits have been discovered on the Tonga-Kermadec Ridge by Hein *et al.* (1988b) but these also do not appear to be associated with polymetallic sulphide deposits.

Another island arc setting in the south-western Pacific where hydrothermal deposits have been recorded is in the New Hebrides Arc in the vicinity of the island of Epi (Exon and Cronan, 1983; Green and Exon, 1988). There are three submarine volcanoes off the island of Epi (Figure 8.16) associated with a submarine caldera, of which the largest was active in 1984 when plumes of fluid and gas were seen to be venting near the northern crater wall. Thus, this volcano must have a high potential for hosting submarine hydrothermal mineral deposits. Exon and Cronan (1983) reported ferruginous sediments containing more than 20% iron and enriched in several other elements of hydrothermal origin at shallow depth near the volcanic cones in the Epi caldera. While of no economic value in themselves, they do point to the potential occurrence of higher grade more valuable deposits in their vicinity.

An additional setting in the Vanuatu area where hydrothermal mineral deposits are likely to occur is in the troughs which are discontinuous along the New Hebrides back-arc between the arc trench system and the North Fiji Basin (Figure 3.12). This has been

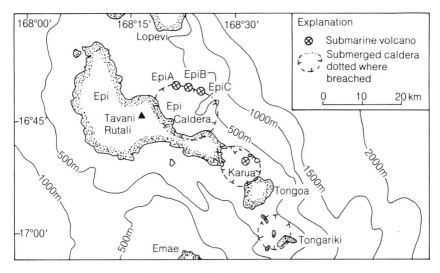

Fig. 8.16 Submarine volcanoes off Epi.

confirmed within the Vot Tande Trough (one of the troughs) by the recent discovery of hydrothermal manganese encrustations there (Recy, 1987).

8.5 RESOURCE POTENTIAL OF SUBMARINE HYDROTHERMAL MINERAL DEPOSITS

It is only in the Red Sea that detailed studies on the resource potential of submarine hydrothermal mineral deposits have been conducted, because it is only there that sufficient data are available for a meaningful resource estimation to be carried out. The Atlantis II Deep of the Red Sea is a potential ore body for zinc, copper and silver, with gold and cobalt as important subsidiary metals. Lange (1985b), has provided a summary of studies on the resource potential of these deposits, and notes that the Deep covers a total area of approximately 62 km^2 with the hydrothermal deposits ranging in thickness from several metres to about 25 m and having an average thickness of 7 to 11 m. The grain size of the deposits is very small, with 80% below 2 micrometres in size. The water/brine content ranges from 95% in the youngest sediments to about 20% in sediments of around 10 000 years old. The average density is 1.22, and 1.36 for more deeply buried strata. The deposits are remarkably inhomogeneous, being well stratified which results in the metal content varying significantly both vertically and laterally. In order to

Table 8.5. Grades and tonnages of metals in the Atlantis II Deep (from Lange, 1985b)

	Grade	Concentration[a]	Tonnages
Zinc	3.41 kg/m³	2.06%	1,890,000
Copper	0.77 kg/m³	0.46%	425,000
Silver	6.77 g/m³	41 ppm	3,750
Gold	–	0.5 ppm	47
Cobalt	–	59 ppm	5,368
Dry salt-free material	–	–	92×10^6
Metalliferous Sediments	–	–	696×10^6

Note:
[a] Based on dry salt-free material.

carry out a resource assessment of these deposits, 628 sediment cores totalling an aggregate 4 km in length were recovered. Geostatistical methods have been employed to assess the metal grades and tonnages for individual blocks in each area of the deposit. A summary of the grades and tonnages of metals based on these data is given in Table 8.5. In terms of total tonnages, the hydrothermal deposits in the Red Sea rival the largest volcanogenic massive sulphide deposits on land, and amounted to a value of about $5 billion at 1985 prices, assuming a recovery typical for on-land volcanogenic massive sulphides.

Other deeps in the Red Sea have not been studied in the same amount of detail as the Atlantis II Deep because their resource potential was recognized at an early stage to be less than that of the Atlantis II Deep. However, hydrothermal deposits occur in several deeps including the Thetis Deep, Gypsum Deep, Nereus Deep, Vema Deep, Erba Deep and Shagara Deep (Bignell *et al.*, 1976b). Gypsum Deep contains iron oxide and iron sulphide-rich sediments as well, of course, as containing gypsum. Vema Deep contains layers of limonitic mud. Nereus Deep contains brine pools below which there are predominantly normal Red Sea sediments intermixed with varying amounts of amorphous oxide and sulphide mud (Bignell *et al.*, 1974). The most metal-rich deposits in the Nereus Deep reported to date are found within the south-east corner of the Deep and appear to be unassociated with present-day brine pools. Sediments from this part of the Deep are similar to the manganite facies sediments in the Atlantis II Deep (Bignell *et al.*, 1976b). Other than in the Atlantis II Deep, some of the most highly metalliferous sediments found in the Red Sea occur in Thetis Deep. The most recently deposited sediments in Thetis Deep are similar to the manganite facies of the Atlantis II Deep. Below these deposits are iron-rich sediments. Minor amounts of sulphides occur in the upper manganite layer. Sediments similar to those found below the brine pool in the Nereus Deep are found in Suakin Deep (Bignell *et al.*, 1976b).

Much less detailed information on the nature and extent of hydrothermal deposits elsewhere on the sea floor is available than for deposits in the Red Sea. The main reason for this is that mid-ocean ridges, convergent plate margins and mid-plate hotspots have only been fragmentally explored, and few, if any, reliable estimates are available of the thickness of the hydrothermal deposits present there, data of great importance in assessing their resource potential in view of the internal inhomogenity of chimneys and mounds discussed earlier. For these reasons there are no detailed estimates of the resource potential of submarine hydrothermal mineral deposits other than in the Red Sea, although general estimates of tonnages have been attempted (see p. 140).

Clark (1989) points out that none of the seafloor polymetallic sulphide deposits discovered to date are as large as the largest polymetallic sulphide deposits on land, which are up to 125 million tonnes in the case of the Kuroko-type deposits of Japan. He poses the question 'if seafloor polymetallic sulphide deposits are modern equivalents of ancient massive sulphide ore deposits, where are the big ones?' He points out that they would not be expected on mid-ocean ridges, citing in support the observations of Converse *et al.* (1984) that most of the hydrothermal mineralization in such settings is lost to the water column in the hydrothermal plume. Rather, in agreement with previous workers, he considers 'geological trap' areas to be favourable sites for large volume sulphide accumulations, where the sulphides are trapped or ponded by some means. As mentioned earlier, such settings include seamount calderas in young oceanic crust, fault scarps bordering axial valleys associated with spreading centres, and sediment buried spreading centres. The last of these was thought by Clark (1989) to have the greatest potential to host the largest polymetallic sulphide deposits. This is because they are areas where: (a) maximum amounts of sulphides will be precipitated due to interaction with and entrapment by the sediment column including contained organic material; and (b) precipitation will take place over the longest period of time due to the insulating effect of the sediments. In agreement with earlier workers, Clark (1989) considers that the western Pacific convergent plate boundaries are likely to provide the most favourable settings for large and rich polymetallic sulphide deposits, not least in regard to the latter in their potential gold content. The western Pacific margin is an area rich in epithermal gold deposits, and this is held by Clark to imply that the submarine polymetallic sulphides in that region are likely to be among the richest in gold of all those on the sea floor.

A number of factors will need to be taken into account when the resource potential of seafloor sulphide deposits comes to be assessed in detail. These will, of course, include the size of the deposit, its grade including the presence or otherwise of rare and normally rare elements which might lead to value enhancement, its mineability, and its ownership. Many seafloor sulphide deposits are comparable with massive sulphides on land (in terms of size) but from the resource point of view simple comparisons with grades and tonnages are meaningless because of the as yet unknown problems which have to be solved in actually mining the deposits.

There is no shortage of massive sulphide deposits on land, and thus mining seafloor sulphides for normal massive sulphide associated elements such as copper and zinc is unlikely within the foreseeable future unless large deposits are found. However, the elevated concentrations of noble metals, particularly gold, in some deposits has maintained a resource interest in them, and, as mentioned earlier, it is widely considered that only a significant enrichment of one or more of the noble metals will make a seafloor massive sulphide deposit worth mining in the near future. The accumulated evidence suggests that, after the Red Sea, this is most likely to take place in a convergent plate margin setting.

References

Agiorgitis, G. and Gundlach, H. (1978) Platin-gehalte in tiefsee manganknollen. *Naturwissenschaften*, **65**, 534.

Aleva, G.J.J. (1973) Aspects of the historical and physical geology of the Sunda Shelf essential for the exploration of submarine tin placers, *Geol. en Mi jnb* **52**, 79–91.

Andrews, H. and Friedrich, G. (1980) Distribution patterns of manganese nodule deposits in the northeast equatorial Pacific, in, *Geology and Geochemistry of Manganese*, **3** (ed. I.M. Varentsov and Gy Grasselly) Akademiai Kiado, Budapest, pp. 107–36.

Aplin, A.C. (1983) The geochemistry and environment of deposition of some ferromanganese deposits from the south equatorial Pacific, Ph.D. thesis, University of London.

Aplin, A.C. and Cronan, D.S. (1985a) Ferromanganese oxide deposits from the Central Pacific Ocean I Ferromanganese oxide encrustations from the Line Islands, *Geochim. Cosmochim. Acta.* **49**, 427–36.

Aplin, A.C. and Cronan, D.S. (1985b) Ferromanganese oxide deposits from the Central Pacific Ocean II Nodules and associated sediments, *Geochim. Cosmochim. Acta.* **49**, 437–51.

Archer, A. (1979) Resources and potential reserves of nickel and copper in manganese nodules, in, *Manganese Nodules: Dimensions and Perspectives*, D. Riedel, Dordrecht, pp. 71–81.

Archer, A. (1987) Sources of confusion: what are marine mineral resources? in, *Marine Minerals* (ed. P.G. Teleki, M.R. Dobson, J.R. Moore and U. von Stackelberg) D. Reidel, Dordrecht, pp. 421–32.

Ardus, D.A. and Harrison, D.J. (1989) The assessment of aggregate resources from the United Kingdom continental shelf, Proceedings EEZ Resources: Technology Assessment Conference IOTC, Honolulu, 1989, p. 5–1–20.

Auzende, J.M., Eissen, J.P., Lafoy *et al.* (1988) Sea floor spreading in the North Fiji Basin (south-west Pacific), *Tectonophysics*, **146**, 317–51.

Auzende, J.M., Urabe, T. and Shipboard Party (1989) Preliminary results of the STARMER I Cruise of the submersible Nautile in the North Fiji Basin. Abs. Joint CCOP/SOPAC – IOC Fourth International Workshop on Geology, Geophysics and Mineral Resources of the South Pacific, 24 September – 1 October 1989, Canberra, Australia, CCOP/SOPAC Misc. Rept. 79, Suva, Fiji, pp. 12–13.

Backer, H. (1973) Rezente hydrothermal-sedimentare lagerstatten-bildung, *Erzmetall* **26**, 544–55.

Backer, H. and Lange, J. (1987) Recent hydrothermal metal accumulation: products and conditions of formation, in, *Marine Minerals* (ed. P.G. Teleki, M.R. Dobson, J.R. Moore and U. von Stackelberg) D. Riedel, Dordrecht, pp. 317–38.

Backer, H, and Richter, H. (1973) Die rezente hydrothermal – sedimentare lagerstatte im Roten Meer, *Geol. Rund.* **62**, 697–740.

Baker, K. and Burnett, W.C. (1988) Distribution texture and composition of modern phosphate pellets in Peru shelf muds, *Marine Geology*, **80**, 195–214.

Ballard, R., Choukrone, P., Cheminee, J.L. *et al.* (1983) Intense hydrothermal activity of the axis of the East Pacific Rise near 13°N: submersible witnesses the growth of sulfide chimney, *Mar. Geophys Res.* **6**, 1–14.

Ballard, R. and Francheteau, J. (1982) The relationship between active sulphide deposition and the axial processes of the Mid-Ocean Ridge, in *Marine Mineral Deposits.* (ed. P. Halbach and P. Winter), Verlag Gluckauf Essen, pp. 137–76.

Bastien-Thiery, H., Lenoble, J.P. and Rogel, P. (1977) French exploration seeks to define mineable nodule tonnages on Pacific floor, *Engng. Min., Journal*, **178**, 86–7.

Baturin, G.N. (1982) *Phosphorites on the Sea Floor*, Elsevier, Amsterdam.

Beerbower, J.R. (1968) *Search for the Past* (2nd Ed) Prentice Hall, Englewood Cliffs, New Jersey.

Beiersdorf, H., Kudrass, H.R. and von Stackelberg, U. (1980) Placer deposits of ilmenite and zircon on the Zambezi Shelf, *Geol. Jahrb.* **D.36**, 5–85.

Bender, M., Klinkhammer, G. and Spencer, D. (1977) Manganese in seawater and the marine manganese balance, *Deep Sea Res.*, **24**, 799–812.

Bertine, K.J. and Keen, J. (1975) A hydrothermal barite from the Lau Basin, *Science*, **188**, 150–2.

Bezrukov, P.L. and Baturin, G.N. (1976) Lithology of oceanic phosphorites, in, *Lithology of Phosphorite Bearing Deposits*, Nauka, Moscow, 20–8.

Bezrukov, P.L. and Skornyakova, N.S. (1976) Manganese nodules from polygons in the S.W. Pacific. *Am. Assoc. Petrol. Geol. Memoirs*, **25**, 376–81.

Bignell, R.D. (1975) Timing, distribution and origin of submarine mineralisation in the Red Sea, *Trans. Inst. Min. Met. B.* **84**, 1–6.

Bignell, R.D. (1978) Genesis of Red Sea metalliferous sediments, *Mar. Mining*, **1**, 212–20.

Bignell, R.D., Cronan, D.S. and Tooms, J.S. (1976a) Metal dispersion in the Red Sea as an aid to marine geochemical exploration, *Trans. Inst. Min. Metall. B.*, **85**, 273–278.

Bignell, R.D., Cronan, D.S. and Tooms, J.S. (1976b) Red Sea metalliferous brine precipitates, *Trans. Geol. Assoc. Canada. Spec. Paper* **14**. 148–84.

Bignell, R.D., Tooms, J.S., Cronan, D.S. and Horowitz, A. (1974) An additional location of metalliferous sediments in the Red Sea, *Nature*, **248**, 127–8.

Biliki, N. (1988) Nearshore Minerals: Solomon Islands, in *Inshore and Nearshore Minerals Workshop Papers*, CCOP/SOPAC, Suva, September (unpublished).

Binns, R.A., Scott, S.D. and PACLARK Participants (1987) Western Woodlark Basin: potential analogue setting for volcanogenic massive sulphide deposits, in *Proc. Pac. Rim Congress 1987, Australian Inst. Min. Metall.*, 531–35.

Binns, R.A., Scott, S.D. and PACLARK Participants (1989) Propagation of sea floor spreading into continental crust, western Woodlark Basin, Papua New Guinea. Abs. Joint CCOP/SOPAC – IOC Fourth International Workshop on Geology, Geophysics and Mineral Resources of the South Pacific, 24 September – 1 October 1989, Canberra, Australia, CCOP/SOPAC Misc. Rept. 79, Suva, Fiji, pp. 14–15.

Binns, R.A., Scott, S.D. and PACLARK Participants (1990) Tectonic, magmatic and hydrothermal activity in the western Woodlark Basin, Papua New Guinea: a propagating marginal basin, *EOS Trans. Am. Geophys. Union*, **71**, 954 Abstract.

Bolton, B.R., Both, R., Exon, N.F. *et al.* (1988) Geochemistry and mineralogy of seafloor hydrothermal and hydrogenetic Mn oxide deposits from the Manus Basin and Bismarck Archipelago region of the south-west Pacific Ocean, *Marine Geology*, **85**, 65–87.

Both, R., Crook, K., Taylor, B. *et al.* (1986) Hydrothermal chimneys and associated fauna in the Manus back-arc basin, Papua New Guinea, *EOS, Trans. Am. Geophys. Union*, **67**, 489–90.

Broadus, J.M. (1987) Seabed materials, *Science*, **235**, 853–60.

Bruland, K.W. (1980) Oceanographic distribution of cadmium, zinc, nickel and copper in the North Pacific, *Earth Planet Sci. Letters*, **47**, 176–98.

Burnett, W.C. (1977) Geochemistry and origin of phosphorite deposits from off Peru and Chile, *Geol. Soc. Amer. Bull.* **88**, 813–23.

Burnett, W.C. (1987) Open ocean phosphorites: in a class by themselves? in, *Marine Minerals* (eds. P. Teleki, M.R. Dobson, J.R. Moore and U. von Stackelberg), D. Riedel, Dordrecht, pp. 119–34.

Burnett, W.C. (1990). Anoxic marine lakes – an analogue environment for insular phosphorite formation, *Abstract 5th Circum-Pacific Energy and Mineral Resources Conference*, 29 July – 3 August 1990, Honolulu, Hawaii. Circum-Pacific Council for Energy & Mineral Resources, AAPG, Tulsa, p. 30.

Burnett, W.C. and Froelich, P. (1988) The origin of marine phosphorite: the results of the R.V. Robert D. Conrad Cruise 23–06 to the Peru Shelf, *Marine Geology*, **80**, 181–346.

Burnett, W.C. and Lee, A.I.N. (1980) The phosphate supply system of the Pacific Region, *Geo. Journal* **4.5**, 423–36.

Burns, V.M. and Burns, R.G. (1978) Post-depositional metal enrichment processes inside manganese nodules from the north equatorial Pacific, *Earth. Planet. Sci. Letts.* **39**, 341–48.

Buser, W. and Grutter, A. (1956) Uber die Natur der Manganknollen, *Schweiz. miner. petrogr. Mitt.* **36**, 49–62.

Calvert, S.E. and Price, N.B. (1977) Geochemical variation in ferromanganese nodules and associated sediments from the Pacific Ocean, *Mar. Chem.* **5**, 43–74.

Calvert, S.E., Price, N.B., Heath, G.R. and Moore, T.C. Jr. (1978) Relationships between ferromanganese nodule composition and sedimentation in a small survey area of the equatorial Pacific, *Jour. Mar. Res.* **36**, 161–83.

Carleton, C.C. and Philipson, P.W. (1987) Report on a study of the marketing and processing of precious coral products in Taiwan, Japan and Hawaii, Forum Fisheries Agency Report 87/13.

Carranza-Edwards, A., Marquez-Garcia, A. and Moraes de la Garza, E. (1987). Distribucion y caracteristicas fisicas externas de nodulos polymetalicos en el sector central del Pacifico Mexicano, *Bol. de Miner*, **3**, 78–94.

Chester, R. (1990) *Marine Geochemistry*, Unwin Hyman, London.

Clark, A. (1989) Marine mineral resources for the 21st Century, in, *Papers from a Workshop on Marine Mining Technology for the 21st Century*, 4–6 December 1989, East-West Center, Honolulu, Hawaii (unpublished).

Clark, A., Humphrey, P., Johnson, C. and Pak, D. (1985) Cobalt-rich manganese crust potential, OCS Study, US Mineral Management Service Report 85–0006.

Cole, T.G. (1983) Oxygen isotope geothermometry and origin of smectites in the Atlantis II Deep, Red Sea, *Earth. Planet. Sci. Letts.* **66**, 166–76.

Colley, N., Cronan, D.S. and Moorby, S.A. (1979) Some geo-chemical and mineralogical studies on newly collected ferro-manganese oxide deposits from the north-west Indian Ocean, in, *Sur La Genèse des Nodules de Manganese* (ed. C. Lalou) Colloque International du CNRS No. **289**, 13–21.

Converse, D.R., Holland, H.D. and Edmond, J. (1984) Flow rates in the axial hot springs of the East Pacific Rise (21°N): implications for the heat budget and the formation of massive sulphide deposits, *Earth. Planet. Sci. Letts.* **69**, 159–75.

Coward, R.N. (1986) A statistical appraisal of regional geochemical data from the South-West Pacific for mineral exploration. PhD thesis, University of London.

Coward, R.N. and Cronan, D.S. (1985) A statistical appraisal of regional geochemical data on marine sediments from the SW Pacific in regard to exploration for detrital, bedrock, phosphatic and hydrothermal mineral deposits, CCOP/SOPAC Tech. Rept. 52, Suva, Fiji.

Coward, R.N. and Cronan, D.S. (1987) A geostatistical evaluation of geochemical data in regard to bedrock and placer mineral exploration in the SW Pacific, *Marine Mining*, **6**, 205–21.

Craig, H. and Poreda, R. (1987) Studies of methane and helium in hydrothermal vent plumes, spreading axis basalts, and volcanic island lavas and gases in southwest Pacific marginal basins, SIO ref. 87–14, Scripps Institute of Oceanography, University of California.

Cronan, D.S. (1967) The geochemistry of some manganese nodules and associated pelagic deposits, Ph.D. thesis, University of London.

Cronan, D.S. (1975) Manganese nodules and other ferromanganese oxide deposits from the Atlantic Ocean, *Jour. Geophys. Res.* **80**, 3831–7.

Cronan, D.S. (1977) Deep sea nodules: distribution and geo-chemistry, in, *Marine Manganese Deposits* (ed. G.P. Glasby) Elsevier, Amsterdam, 11–44.

Cronan, D.S. (1978) Manganese nodules, controversy upon controversy, *Endeavour* **2**, 80–3.

Cronan, D.S. (1980) *Underwater Minerals*, Academic Press, London.

Cronan, D.S. (1983) Metalliferous sediments in the CCOP/SOPAC region of the southwest Pacific, with particular reference to geochemical exploration for the deposits, CCOP/SOPAC Tech. Bull. 4, Suva, Fiji.

Cronan, D.S. (1984) Criteria for the recognition of areas of potentially economic manganese nodules and encrustations in the CCOP/SOPAC region of the central and southwestern tropical Pacific, *South Pac. Mar. Geol. Notes*, **3**, 1–17.

Cronan, D.S. (1985) A wealth of sea floor minerals, *New Scientist*, 6 June.

Cronan, D.S. (1989a) Overview of mineral resources in the EEZ, in Proc. EEZ Resources Technology Assessment Conference, IOTC, Honolulu.

Cronan, D.S. (1989b) Hydrothermal metalliferous sediments in the SW Pacific, in, *Oceanography, 1988* (eds. A. Ayala Castanares, W.S. Wooster and A. Yanez-Arancibia) UNAM Press, Mexico, pp. 149–66

Cronan, D.S. and Hodkinson, R. (1989) Manganese nodules and cobalt-rich crusts in the EEZ's of the Cook Islands, Kiribati and Tuvalu, III, Nodules and crusts in the EEZ of western Kiribati (Phoenix and Gilbert Islands), CCOP/SOPAC Tech. Rept. 100, Suva, Fiji.

Cronan, D.S. and Hodkinson, R. (1990) Manganese nodules and cobalt-rich crusts in the EEZs of the Cook Islands, Kiribati and Tuvalu, IV, Nodules and crusts in the EEZ of Tuvalu (Ellice Islands) CCOP/SOPAC Tech. Rept. 102, Suva, Fiji.

Cronan, D.S. and Tooms, J.S. (1967a) Sub-surface concentrations of Mn nodules in Pacific sediments, *Deep Sea Res.* **14**, 117–19.

Cronan, D.S. and Tooms, J.S. (1967b) Geochemistry of manganese nodules from the NW Indian Ocean, *Deep-Sea Res.* **14**, 239–49.

Cronan, D.S. and Tooms, J.S. (1968) A microscopic and electron probe investigation of Mn nodules from the NW Indian Ocean, *Deep Sea Res.* **15**, 215–23.

Cronan, D.S. and Tooms, J.S. (1969) The geochemistry of Mn nodules and associated pelagic deposits from the Pacific and Indian Oceans, *Deep Sea Res.* **16**, 335–59.

Cronan, D.S., Glasby, G.P., Moorby, S.A. *et al.* (1982) A submarine hydrothermal manganese deposit from the south-west Pacific island arc, *Nature,* **298**, 456–8.

Cronan, D.S., Hodkinson, R. and Miller, S. (1991) Manganese nodules in the EEZs of island countries in the southwestern Pacific, *Marine Geology*, **98**, 425–35.

Cronan, D.S., Hodkinson, R., Harkness, D.D. *et al.* (1986) Accumulation rates of hydrothermal metalliferous sediments in the Lau Basin, S.W. Pacific. *GeoMarine Letters*, **6**, 51–6.

Cronan, D.S., Hodkinson, R., Miller, S. and Hong, L. (1989) Manganese nodules and cobalt-rich crusts in the EEZs of the Cook Islands, Kiribati and Tuvalu. II Nodules and crusts in the EEZs of the Cook Islands and part of eastern Kiribati (Line Islands), CCOP/SOPAC Tech. Rept. 99, Suva. Fiji.

Cronan, D.S., Tiffin, D. and Meadows, P. (1987) A study of manganese nodules, crusts and deep-sea sediments in the northern Cook Islands, central Line Islands and adjacent high seas. Cruise report of the Crossgrain Expedition, Leg. 3, Scripps Institute of Oceanography, Misc. cruise reports.

Cullen, D.J. (1979) Mining minerals from the sea floor: Chatham Rise phosphorite, *New Zealand Agricult. Sci.* **13**, 85–91.

Cullen, D.J. (1986) Submarine phosphatic sediments of the SW Pacific, in, *Sedimentation and Mineral Deposits in the Southwestern Pacific Ocean* (ed. D.S. Cronan), Academic Press, London, pp. 183–236.

Cullen, D.J. and Burnett, W.C. (1986) Phosphorite associations on seamounts in the tropical S.W. Pacific Ocean. *Marine Geology*, **71**, 215–36.

Cullen, D.J. and Burnett, W.C. (1987) Insular phosphorite on submerged atolls in the tropical southwest Pacific, *Search*. **18**, 311.

Cullen, D.J., Kudrass H. and von Rad, U. (1981) Preliminary results of the 1981 Sonne investigation of Chatham Rise phosphorite deposits east of New Zealand, Proc. Inter Ocean 1981 Conference, Dusseldorf, 301/1–6.

De Carlo, E.H., McMurtry, G.M. and Yeh, H.W. (1983) Geochemistry of hydrothermal deposits from Loihi submarine volcano, Hawaii. *Earth. Planet. Sci. Letts.* **66**, 438–49.

Dehais, J.A. and Wallace, W.A. (1988) Economic aspects of offshore sand and gravel mining, *Marine Mining*, **7**, 35–48.

Dunham, K.C. (1969) Practical geology and the natural environment of man, II, seas and oceans, *Q.J. Geol. Soc. Lond.* **124**, 101–29.

Eade, J.V. (1980) Review of precious coral in CCOP/SOPAC member countries, CCOP/SOPAC Tech. Rept. 8, Suva, Fiji.

Edmond, J.M., Measures, C., Magnum, B. *et al.* (1979) On the formation of metal rich deposits at ridge crests, *Earth. Planet. Sci. Letts.* **46**, 19–30.

Edmond, J.M, von Damm, K.L., McDuff, R.E. and Measures, C.I. (1982) Chemistry of hot springs on the East Pacific Rise and their effluent dispersal, *Nature*, **297**, 187–91.

Elderfield, H. (1976) Hydrogenous material in marine sediments, in,

Chemical Oceanography 5 (eds. J.P. Riley and R. Chester), Academic Press. London, pp. 137–216.

Emery, K. and Noakes, L.C. (1968) Economic Placer Deposits of the Continental Shelf, *ECAFE CCOP Tech. Bull.*, **1**, CCOP, Bangkok.

Exon, N.F. (1981a) The magnetite sands of Vanuatu, in *Report on the Inshore and Nearshore Resources Training Workshop, Suva, Fiji, July 1981*, CCOP/SOPAC, Suva, pp. 29–30.

Exon, N.F. (1981b) Cruise Report, Vanuatu Offshore Survey, Precious Corals and Metalliferous Sediments VA-80(3), Cruise Report No. 48. CCOP/SOPAC, Suva.

Exon, N.F. and Cronan, D.S. (1983) Hydrothermal iron deposits and associated sediments from submarine volcanoes off Vanuatu, Southwest Pacific, *Marine Geology*, **52**, 433–52.

Falconer, R.K.H. (1989) Chatham Rise phosphates: a deposit whose time has come (and gone), *Marine Mining* **8**, 55–67.

Fouquet, Y., Auclair., G., Cambon, P. and Etoubleau, J. (1988) Geological setting and mineralogical and chemical investigations on sulphide deposits near 13°N on the East Pacific Rise, *Marine Geology*, **84**, 145–78.

Fouquet, Y. and Shipboard Party (1989) High temperature poly-metallic sulphide deposits in back-arc environments: diving with 'Nautile' in the Lau Basin, Abs. Joint CCOP/SOPAC – IOC Fourth International Workshop on Geology, Geophysics and Mineral Resources of the South Pacific, 24 September – 1 October 1989, CCOP/SOPAC Misc. Rept. 79, Suva, Fiji, 43.

Francheteau, J. and Ballard, R. (1983) The East Pacific Rise near 21°N: inference for along strike variability of axial processes of the Mid-Ocean Ridge, *Earth. Planet. Sci. Letts.* **64**, 93–116.

Francis, T.G. (1985) Resistivity measurements of an ocean floor sulphide mineral deposit from the submersible Cyana, *Mar. Geophys. Res.* **7**, 419–38.

Friedrich, G. and Pluger, W. (1974) Die verteilung von mangan, eisen, kobalt, nickel, kupfer und zink in manganknollen verschiedener felder, *Meersestecknik* **5**, 203–6.

Froelich, P.N., Arthur, M.A., Burnett, W.C. *et al.* (1988) Early diagenesis of organic matter in Peru continental margin sediments: phosphorite precipitation. *Mar. Geol.*, **80**, 309–42.

Gass, I.C. (1977) The age and extent of the Red Sea oceanic crust, *Nature*, **265**, 722–3.

Girdler, R.W. and Styles, P. (1974) Two stages of Red Sea floor spreading, *Nature*, **247**, 7–11.

Glasby, G.P. (1986) Neashore mineral deposits in the SW Pacific, in,

Sedimentation and Mineral Deposits in the Southwestern Pacific Ocean (ed. D.S. Cronan), Academic Press, London pp. 149–81.

Goddard, D.A., Thompson, C., Jones, E. and Okada, H. (1987) The chemistry and mineralogy of ferromanganese encrustations on rocks from the Sierra Leone Rise, equatorial Mid-Atlantic Ridge and New England seamount chain, *Marine Geology*, **77**, 87–98.

Goldberg, E.D. (1954) Marine Geochemistry I: Chemical scavengers of the sea. *J. Geol.*, **62**, 249–65.

Goldberg, E.D. and Arrhenius, G. (1958) Chemistry of Pacific pelagic sediments, *Geochim. et Cosmochim. Acta.* **13**, 153–212.

Goldfarb, M.S., Converse, D.R., Holland, H.D. and Edmond, J. (1983) The genesis of hot spring deposits on the East Pacific Rise at 21°N, *Econ. Geol. Monogr.* **5**, 184–97.

Gooding, K. (1988) Mineral sands reveal key factor of Minorco's bid, *Financial Times*, 1 November 1988.

Gorsline, D.S. (1963) Bottom sediments of the Atlantic shelf and slope of the southern United States, *Jour. Geol.* **71**, 422–40.

Green, D. (1970) Marine Mineral Development, Geol. Survey of Fiji (unpublished report), Mineral Resources Dept., Suva, Fiji.

Green, H.G. and Exon, N.F. (1988) Acoustic stratigraphy and hydrothermal activity within Epi submarine caldera, Vanuatu, New Hebrides Arc. *Geo. Marine Letters*, **8**, 34–48.

Greenslade, J. (1975) Manganese-biota associations in north eastern Equatorial Pacific sediments, Ph.D. dissertation, University of California, San Diego.

Grigg, R.W. (1977) *Hawaii's Precious Corals*, Island Heritage, Honolulu, Hawaii.

Grigg, R.W. and Eade, J.V. (1981) Precious corals, in, *Report on the Inshore and Nearshore Resources Training Workshop, Suva, Fiji, July 1981.* CCOP/SOPAC, Suva, pp. 13–18.

Griggs, A.B. (1945) Chromite bearing sands of the southern part of the coast of Oregon, *US Geol. Survey Bull*, 945–E, 113–50.

Halbach, P. (1989) Jade hydrothermal field, Okinawa Trough: first discovery of Ag and Au rich massive sulphides in an intra-continental back-arc basin, Abs, 20th Underwater Mining Institute, Madison, Wisc. October 1989, Univ. Wisc Sea Grant Office.

Halbach, P. and Ozkara, M. (1979) Morphological and geochemical classification of deep sea ferromanganese nodules and its genetical interpretation, in *La Genese des Nodules de Manganese* (ed. C. Lalou), Colloques Internationaux du CNRS, 289, pp. 77–89.

Halbach, P. and Puteanus, D. (1984) The influence of the carbonate dissolution rate on the growth and composition of Co-rich ferromanganese crusts from Central Pacific seamount areas, *Earth. Planet. Sci. Letts.* **68**, 73–87.

Halbach, P. and Puteanus, D. (1985) Cobalt-rich ferromanganese deposits within the Johnson Island EEZ – Environmental and Resource Data, Report to US Minerals Management Service (unpublished).

Halbach, P., Manheim, F.T. and Otten, P. (1982) Co-rich ferro-manganese deposits in the marginal seamount region of the Central Pacific Basin – results of the Mid Pac. '81, *Erzmetall*, **35**, 447–53.

Halbach, P., Puteanus, D. and Manheim, F. (1984) Platinum concentrations in ferromanganese seamount crusts from the Central Pacific, *Naturwissenschaften*, **71**, 577–9.

Halbach, P., Sattler, C.D., Teichmann, F. and Wahsner, M. (1989) Cobalt-rich and platinum bearing manganese crust deposits on seamounts: nature, formation and metal potential, *Marine Mining*, **8**, 23–40.

Halbach, P., Segl, M., Puteanus, D. and Mangini, A. (1983) Relationships between Co-fluxes and growth rates in ferro-manganese deposits from Central Pacific seamount areas, *Nature*, **304**, 716–19.

Halbach, P., Ko-ichi Nakamura, Wahsner, M. *et al.* (1989) Probable modern analogue of Kuroko-type massive sulphide deposits in the Okinawa Trough back-arc basin, *Nature*, **338**, 496–9.

Hale, P. (1988) Global overview of exploration and mining activities. *Inshore and Nearshore Minerals Workshop Papers*, CCOP/SOPAC, Suva, September 1988 (unpublished).

Hannington, M.D. and Scott, S.D. (1988) Gold and silver potential of polymetallic sulphide deposits on the sea floor, *Marine Mining* **7**, 271–85.

Harper, J.R. (1988) Precious coral prospecting strategies for the South Pacific region, CCOP/SOPAC Tech. Rept. 84, CCOP/SOPAC, Suva.

Harrison, D.J. and Ardus, D.A. (1990) Geological investigations for marine aggregates offshore East Anglia, *J. Soc. Underwater Technol.* **16**, 9–14.

Hartman, M. (1985) Atlantis II Deep geothermal brine system. Chemical processes between hydrothermal brines and Red Sea deep water, *Marine Geology*, **64**, 157–77.

Hawkins, J. and Helu, S. (1986) Polymetallic sulphide deposit from black smoker chimney: Lau Basin, EOS, *Trans. Am. Geophys. Union* **67**, p. 378.

Haymon, R.M. and Kastner, M. (1981) Hot spring deposits on the East Pacific Rise at 21°N: preliminary description of mineralogy and genesis, *Earth. Planet. Sci. Letts.* **53**, 363–81.

Haynes, B.W., Law, S.L. and Barron, D.C. (1982) Mineralogical and elemental description of Pacific manganese nodules, U.S. Bureau of Mines Information Circular 8906.

Heezen, B.C., Matthews, J.L., Catalano, R. *et al.* (1973) Western Pacific Guyots, Init. Rept. DSDP Leg 20, US Government Printing Office, Washington, DC, pp. 653–702.

Hein, J.R., Gein, L.M. and Morrison, M.S. (1988a) Submarine ferromanganese mineralisation in active volcanic arc systems, in, *Proceedings of PACON 88, Pac. Cong. on Marine Sci. and Tech.*, Honolulu, Hawaii, May, 83–8.

Hein, J.R., Morrison, M.S. and Gein, L.A. (1991) Central Pacific cobalt-rich ferromanganese crusts: historical perspective & regional variability, in, *Geology and Offshore Mineral Resources of the Central Pacific Basin.* (B. Keating and B. Bolton eds.), Circum-Pacific Council for Energy and Mineral Resources, Earth Science Series, Houston, Texas (in press).

Hein, J.R., Schwab, W.C. and Davis, A.S. (1988b) Cobalt and platinum rich ferromanganese crusts and associated substrate rocks from the Marshall Islands, *Marine Geology* **78**, 255–83.

Hein, J.R., Manheim, F.T., Schwab, W.B. and Davis, A.S. (1985a) Ferromanganese crusts from Necker Ridge, Horizon Guyot and S.P. Lee Guyot: geological considerations, *Marine Geology*, **69**, 25–54.

Hein, J.R., Morgenson, L.A., Clague, D.A. and Koski, R.A. (1987) Cobalt-rich ferromanganese crusts from the Exclusive Economic Zone of the United States and nodules from the oceanic Pacific, in, *Geology and Resource Potential of the Continental Margin of Western North America and Adjacent Ocean Basins*, Circum-Pacific Council for Energy & Mineral Resources, Earth Science Series 6, Houston, Texas, 753–71.

Hein, J.R., Manheim, F.T., Schwab, W.C. *et al.* (1985b) Geological and geochemical data for seamount and associated ferromanganese crusts in and near the Hawaiian, Johnson Island and Palmyra Island Exclusive Economic Zones, USGS Open File Report, 85–292.

Hekinian, R. and Bideau, D. (1985) Volcanism and mineralisation of the oceanic crust on the East Pacific Rise, in, *Metallogeny of Basic and Ultrabasic Rocks*, Inst. Min. Metal. London, pp. 1–20.

Hekinian, R., Fevrier, M., Bischoff, J.L. *et al.* (1980) Sulphide deposits from the East Pacific Rise near 21°N, *Science,* **207**,

1433–44.

Helsley, C.E., Keating, B., De Carlo, E. *et al.* (1985) Resource assessment of cobalt-rich ferromanganese crusts within the Hawaiian Exclusive Economic Zone, Final Report, 14-12-001-30177, US Minerals Management Service.

Herrouin, G., Lenoble, J.P., Charles, C. and Mauviel, F. (1989) A manganese nodule industrial venture would be profitable: summary of a 4-year study in France, Proc. Offshore Technology Conference, Houston, OTC, 5997, 321–33.

Hodkinson, R. and Cronan, D.S. (1991) Regional variability in cobalt-rich ferromanganese crusts from the central equatorial Pacific, *Marine Geology*, **98**, 437–47.

Holmes, R. (1981) Nearshore magnetite and gold placers in Fiji, in, *Report on the Inshore and Nearshore Resources Training Workshop, Suva, Fiji, July 1981*, CCOP/SOPAC, Suva, pp. 24–5.

Hosking, K. (1971) The offshore tin deposits of southeast Asia, in *Detrital Heavy Minerals*. ECAFE CCOP Tech. Bull. 5, CCOP, Bangkok, 112–29.

Hosking, K. and Ong, P. (1963) The distribution of tin and certain other heavy minerals in the superficial portions of the Gwithian/Hayle beach of West Cornwall, *Trans. Roy. Geol. Soc. Cornwall*, **19**, 351–90.

Humphrey, P. (1988) Equatorial enrichment of Pacific seamount ferromanganese crusts rich in cobalt, in, *Proceedings of PACON 88 Pac. Cong. on Marine Sci. & Tech.*, Honolulu, Hawaii, May, 1988.

Jones, H.A., and Davies, P.J. (1979) Preliminary studies of offshore placer deposits, Eastern Australia, *Marine Geology*, **30**, 243–68.

Jones, C.J. and Murray, J.W. (1985) The geochemistry of manganese in the northeast Pacific Ocean off Washington, *Limnol. Oceanogr.*, **30**, 81–92.

Jones, H.A., Kudrass, H.R., Schlüter, H.U. and von Stackelberg, U. (1982) Geological and geophysical work on the east Australian shelf between Newcastle and Frazer Island – a summary of results from SONNE Cruise SO-15, *Geol. Jahrb.* **D-56**, 197–207.

Kang, J.K., Hein, J.R., Park, B.K. and Yoon, S.H. (1990) Preliminary results of KORDI-USGS cooperative cruise for ferromanganese crusts in the western Marshall Islands, Abs. 5th *Circum Pacific Energy & Mineral Resources Conference*, 29 July – 3 August 1990, Honolulu, Hawaii.

Kaufman, R. (1976) Offshore hard mineral resource potential and mining technology, in, *Symposium on Marine Resource Development in the Middle Atlantic States: Soc. Naval Architects and Marine Engineering, Chesapeake and Hampton Roads Section*.

Keating, B.H. (1989) Seamount morphology and manganese crust resources in the Central Pacific Basin, Abs. Joint CCOP/SOPAC-IOC Fourth International Workshop on Geology, Geophysics & Mineral Resources of the South Pacific. Canberra, Australia, 24 September – October 1989, CCOP/SOPAC Misc. Rept. 79. Suva, Fiji.

Kildow, J.T., Bever, M.B., Dar, V.K. and Capstaff, A.E. (1976) Assessment of economic and regulatory conditions affecting ocean mineral resource development, MIT unpublished report for US Interior Dept., Ocean Mining Admin.

Kitekei'ako, F. and Harper, J.R. (1988) High resolution seismic survey for lagoonal sand and gravel resources in Tongatapu, Kingdom of Tonga, in, *Inshore and Nearshore Minerals Workshop Papers*, CCOP/SOPAC, Suva, September (unpublished).

Klinkhammer, G. and Bender, M.L. (1980) The distribution of manganese in the Pacific Ocean, *Earth. Planet. Sci. Letts.* **46**, 361–84.

Komar, P.D. and Wang, C. (1984) Processes of selective grain transport and the formation of beach placers, *Jour. Geol.* **92**, 637–55.

Koski, R.A. (1989) New data from the Escanaba Trough: implications for other sediment covered ridge axes and Besshi-type sulfides on land, Abstract, 20th Underwater Mining Institute, Madison, Wisc. October 1989. University of Wisconsin Sea Grant Office.

Koski, R.A., Shanks, W.C., Bohrson, W.A. and Oscarson, R.L., (1988) The composition of massive sulphide deposits from the sediment covered floor of Escanaba Trough, Gorda Ridge: implications for depositional processes, *Can. Mineral.* **26**, 655–73.

Koski, R.A. Lonsdale, P.F., Shanks, W.C. *et al.* (1985) Mineralogy and geochemistry of a sediment-hosted hydrothermal sulphide deposit from the southern trough of Guaymas Basin, Gulf of California, *Jour. Geophys. Res.* **90**, 6695–707.

Ku, T.L. (1977) Rates of accretion, in, *Marine Manganese Deposits* (ed. G.P. Glasby) Elsevier, Amsterdam, 249–68.

Kudrass, H. (1987) Sedimentary models to estimate the heavy mineral potential of shelf sediments, in, *Marine Minerals* (eds. P.G. Teleki, M.R. Dobson, J.R. Moore and U. von Stackelberg), D. Riedel, Dordrecht, pp. 39–56.

Kudrass, H. and Cullen, D.J. (1982) Submarine phosphorite nodules from the central Chatham Rise off New Zealand – composition, distribution and reserves, *Geol. Jaharb.* **D.51**, 3–41.

Kulm, L.D. (1988) Potential heavy mineral and metal placers on the southern Oregon continental shelf, *Marine Mining*, **7**, 361–95.

Lalou, C., Brichet, E. and Hekinian, R. (1985) Age dating of sulphide deposits from axial and off axial structures on the East Pacific Rise near 12°50'N, *Earth Planet Sci. Letts*. **75**, 59–71.

Lange, J. (1985a) Massive sulphide deposits at recent sea floor spreading centres: genesis, composition and distribution, in *Proc. Pac. Mar. Minerals Training Course, Honolulu*, June *4–28 1985* (ed. C. Johnson and A. Clark), East-west Centre, Honolulu, pp. 91–116.

Lange, J. (1985b) Hydrothermal metalliferous sediments at mid-ocean ridges and the Red Sea axial trough: genesis and composition, in *Proc. Pac. Mar. Minerals Training Course, Honolulu, June 4–28 1985* (ed. C. Johnson and A. Clark) East-west Centre, Honolulu, pp. 71–90.

McGregor, B. and Lockwood, M. (1985) *Mapping and Research in the Exclusive Economic Zone*. US Dept, of the Interior and National Oceanic and Atmospheric Administration, Washington, DC.

Mackay, A.D., Gregg, P.E.H. and Seyrs, J.K. (1980) A preliminary evaluation of Chatham Rise phosphorite as a direct-use phosphatic fertilizer, *NZJ. Agricult. Res*. **23**, 441–9.

Malahoff, A. (1982) Geology and chemistry of hydrothermal deposits from active submarine volcano, Loihi, Hawaii, *Nature* **298**, 234–9.

Manheim, F.T. (1974) Red Sea Geochemistry, *Init. Repts. Deep Sea Drilling Proj*. **23** US Government Printing Office, Washington, DC, pp. 975–90.

Manheim, F.T. (1982) Cruise Report, GYRE Cruise 11–82, U.S. Geol. Survey. Office of Marine Geology, Atlantic-Gulf Branch (unpublished).

Manheim, F.T. (1986) Marine cobalt resources, *Science*, **232** 600–8.

Manheim, F.T. Popenoe, P., Siapno, W.D. and Lane, C.M. (1982) Manganese phosphorite deposits of the Blake Plateau (Western North Atlantic), in, *Marine Mineral Deposits: New Research Results and Economic Prospects* (eds. P. Halbach and P. Winter) Verlag Gluckauf, Essen, pp. 9–44.

Martin, J.H. and Knauer, G.A. (1984) VERTEX: manganese transport through the oxygen minima, *Earth. Planet. Sci. Letts*. **67**, 35–47.

Menard, H.W. (1964) *The Marine Geology of the Pacific* McGraw Hill, New York.

Mero, J. (1965) *The Mineral Resources of the Sea*, Elsevier, Amsterdam.

Meylan, M.A. (1974) Field description and classification of

manganese nodules, *Hawaii Inst. Geophys. Rept. HIG-74-9*, 158–68.

Miller, A.R., Densmore, C.D., Degens, E.T. *et al.* (1966) Hot brines and recent iron deposits in deeps of the Red Sea, *Geochim. Cosmochim. Acta.* **30**, 341–59.

Minetti, M. and Bonavia, F.F. (1984) Copper-ore grade hydrothermal mineralisation discovered in a seamount in the Tyrrhenian Sea (Mediterranean): is the mineralisation related to porphyry coppers or to base metal lodes, *Mar. Geol.* **59**, 271–82.

Mizuno, A. and Moritani, T. (1976) Some results of surveys for manganese nodule deposits in the Pacific Ocean by the Geological Survey of Japan, *CCOP/SOPAC Tech. Bull. 2, Suva, Fiji*, pp. 62–79.

Moncrieff, A.G. and Smale-Adams, K.B. (1974) The economics of first generation manganese nodule operations, *Mining Congress Journal*, December, 46–50.

Moorby, S.A. (1978) The geochemistry and mineralogy of some ferromanganese oxides and associated deposits from the Indian and Atlantic Oceans, Ph.D. thesis, University of London.

Moorby, S.A. and Cronan, D.S. (1983) Geochemistry of hydrothermal and pelagic sediments from the Galapagos hydrothermal mounds field, DSDP Leg 70. *Mineralog. Mag.* **47**, 291–300.

Moorby, S.A., Cronan, D.S. and Glasby, G.P. (1984) Geochemistry of hydrothermal Mn-oxide deposits from the SW Pacific island arc, *Geochim. Cosmochim. Acta.* **48**, 433–41.

Moore, J.R. and Welkie, C.J. (1976) Metal bearing sediments of economic interest, coastal Bering Sea, in *Proc. Symp. on Sedimentation*, Geol. Soc. Alaska, K1-K17.

Moore, T.G. Jr. and Heath, G.R. (1966) Manganese nodules, topography and thickness of Quarternary sediments in the Central Pacific, *Nature*, **212**, 983–85.

Murray, J. and Renard, A.F. (1891) *Deep-Sea Deposits*, Report of the Scientific Results of *HMS Challenger*, 1873–76, HMSO.

Murray, L.G. (1969) Exploration and sampling methods employed in the offshore diamond industry. *Proc. Ninth Commonwealth Min. and Metal. Cong.*, **2**, 71–94.

Nunny, R.S. and Chillingworth, P.C.H. (1986) *Marine Dredging for Sand and Gravel*, HMSO, London.

O'Neill, D. and Woolsey (1988) Backreef calcareous aggregates as an industrial mineral resource for Pacific island nations, in, *Inshore and Nearshore Minerals Workshop Papers*, CCOP/SOPAC, Suva, September 1988 (unpublished).

Parrish, F.G. (1988) Management of the UK marine aggregate

dredging, *J. Soc. Underwater Technol.* **14**, 20–5.

Perissoratis, C., Moorby, S.A., Angelopoulos, I. *et al.* (1988) Mineral concentrations in the recent sediments off eastern Macedonia, Northern Greece: geological and geochemical considerations, in, *Mineral Deposits within the European Community* (eds, J. Boissonnas and P. Omenetto) Springer Verlag, Berlin, pp. 530–52.

Perissoratis, C., Moorby, S.A., Papavasiliou, F. *et al.* (1987) The geology and geochemistry of surficial sediments off Thraki, Northern Greece, *Marine Geology* **74**, 209–24.

Peterson, C.D., and Binney, S.E. (1988) Compositional variations of coastal palcers in the Pacific northwest, *Marine Mining*, **7**, 397–416.

Petterson, H. and Rotschi, H. (1952) The nickel content of deep sea deposits, *Geochim. et Cosmochim Acta.* **2**, 81–90.

Pevear, D.R. and Pilkey, O.H. (1966) Phosphorite in Georgia shelf sediments, *Geol. Soc. Amer. Bull.* **77**, 849–58.

Pichocki, C. and Hoffert, M. (1987) Characteristics of Co-rich ferromanganese nodules and crusts sampled in French Polynesia, *Marine Geology*, **77**, 109–19.

Pickard, G.L. (1975) *Descriptive Physical Oceanography*, Pergamon, Oxford.

Piper, D.Z., Gardner, J.V. and Cook, H.E. (1978) Lithic and acoustic stratigraphy of the equatorial North Pacific, DOMES Sites A, B and C, in, *Marine Geology and Oceanography of the Pacific Manganese Nodule Province* (eds. J.L. Bischoff and D.Z. Piper) Plenium, New York, pp. 309–48.

Prabhakar Rao, G. (1968) Sediments of the nearshore region off Neendakara, Kayankulam coast and Ashtamudi and Vatala estuaries, Kerala, India, *Bull. Nat. Inst. Sciences, India*, **38**, 513–51.

Recy, J. (1987) Preliminary results of petrological and mineralogical studies on manganiferous encrustations dredged in New Hebrides (Vanuatu), *Abs. E.U.G. IV, Strasbourg, Terra Cognita*, **7**, 274

Richey, J.L. (1988) Economic reconnaissance of selected placer deposits of the US Exclusive Economic Zone, *Marine Mining*, **7**, 219–31.

Richmond, B.M. (1990) Aggregate resources in reef environments: three examples from the South Pacific, *Abs. 5th Circum. Pacific Energy & Mineral Resources Conf. July 29 – August 3 1990*, Honolulu, Hawaii, Circum. Pacific Council for Energy and Mineral Resources, AAPG, Tulsa, p. 66.

Riggs, S.R. (1984) Palaeoceanographic model of Neogene phosphorite deposition, US Atlantic Continental Margin, *Science*, **223**,

121–3.

Riggs, S.R. (1987) Model of Tertiary phosphorites on the world's continental margins, In, *Marine Minerals* (eds. P. Teleki, M.R. Dobson, J.R. Moore and U. von Stackelberg) D. Riedel, Dordrecht, pp. 99–118.

Riggs, S.R. and Manheim, F.T. (1988) Mineral resources of the US Atlantic continental margin, in, *The Geology of North America* (eds. R.E. Sheridan and J.A. Grow) Vol. 12, The Atlantic Continental Margin. Geol. Soc. Amer. Boulder, pp. 501–20.

Rona, P. (1984) Hydrothermal mineralisation at sea floor spreading centres, *Earth. Sci. Rev.* **20**, 1–104.

Rosales Hoz, L. and Carranza-Edwards, A. (1988) Polymetallic nodule study from the oceanic area near Clarion Island, Colima, Mexico, *Abs. Joint Oceanographic Assembly, Mexico*, p. 94.

Rougerie, F. and Wauthy, B. (1989) Une nouvelle hypothese sur la genese des phosphates d'atolls: le role du processus d'endo-upwelling, *C.R. Acad. Sci. Paris, t.308 Serie* **11**, 1043–7.

Roy, S., Dasgupta, S., Mukhopadhyay, S. and Fukuoka, M. (1990) Atypical ferromanganese nodules from pelagic areas of the Central Indian Basin, equatorial Indian Ocean, *Marine Geology*, **92**, 269–84.

Sanderson, B. (1985) How bioturbation supports manganese nodules at the sediment-water interface, *Deep Sea Res.* **32**, 1281–85.

Scott, S.D. (1987) Seafloor polymetallic sulphides, scientific curiosities or mines of the future, in, *Marine Minerals* (ed. P. Teleki, M.R. Dobson, J.R. Moore and U. von Stackelberg) D. Riedel, Dordrecht, pp. 277–300.

Segl, M., Mangini, A., Bonani, H.J. *et al.* (1984) Be[10] dating the inner structure of Mn-encrustations applying the Zurich Tandem Accelerator, *Nucl. Inst. Meth.* **B5**, 359–64.

Seyfried, W. and Bischoff, J. (1977) Hydrothermal transport of heavy metals by seawater: the role of seawater/basalt ratio, *Earth. Planet. Sci. Letters*, **34**, 67–71.

Sheldon, R.P. (1980) Epsodicity of phosphate deposition and deep ocean circulation, in, *Marine phosphorites – geochemistry, occurrence, genesis, Spec. Publ. Soc. Econ. Paleont. Mineral*, **29**, 239–47. Tulsa.

Siapno, W. (1989) Oral communication, 20th Underwater Mining Institute, Madison, Wisc.

Siddiquie, H.N. and Rajamanickam, G.V. (1979) Surficial mineral deposits of the continental shelf of India, in, *Offshore Mineral Resources*, Documents, BRGM, 7, Orleans, France, pp. 233–58.

Siddiquie, H.N., Rajamanickam, G.V. and Almeida, F. (1979) Offshore ilmenite placers of Ratnagiri, Konkan coast, Maharashtra,

India, *Marine Mining*, **2**, 91–118.

Simoneit, B.R. and Lonsdale, P.F. (1982) Hydrothermal petroleum in mineralised mounds at the seabed of Guaymas Basin, *Nature*, **295**, 198–202.

Sorem, R.K. (1967). Manganese nodules: nature and significance of internal structure, *Econ. Geol.* **62**, 141–7.

Speiss, F.N., MacDonald, K.C., Atwater, T. *et al.* (1980) East Pacific Rise: hot springs and geophysical experiments, *Nature*, **207**, 1421.

State of Hawaii Marine Mining Programme (1987) Mining development scenario for cobalt-rich manganese crusts in the Exclusive Economic Zones of the Hawaiian Archipelago and Johnston Island, *Ocean Resources Branch Contribution* **38**, 326p.

Stoffers, P. and Shipboard Scientific Party (1987) Cruise Report SONNE 47 – Midplate Volcanism, Central South Pacific, French Polynesia, *Repts. Geol. Palaont. Inst., Univ. Kiel. Nr. 19.*

Stride, A.N. (1963) Current-swept sea floors near the southern half of Great Britain, *Quart. J. Geol. Soc. Lond.* **119**, 175–99.

Summerhayes, C.P. (1967) Marine environments of economic mineral deposition around New Zealand: a review, NZ Jl. Mar. Freshwater Res. **1**, 267–82.

Taylor, B. (1979) Bismarck Sea: evolution of a back-arc basin, *Geology*, **7**, 171–74.

Taylor, B. (1984) A geophysical survey of the Woodlark-Solomons region, CCOP/SOPAC Tech. Rept. 324, CCOP/SOPAC, Suva, Fiji.

Taylor, G.R. (1974) Volcanogenic mineralisation in the islands of the Florida group, BSIP, *Trans. Intn. Min. Metall.* **B.83**, 120–30.

Trichet, J. and Rougerie, F. (1990) Atoll phosphate genesis: role of lagoonal biomass and possible role of endo-upwelling processes in phosphate concentration, *Abstract. 5th Circum. Pacific Energy and Mineral Resources Conference*, 29th July–3rd August, Honolulu, Hawaii, p. 74.

Tsurusaki, K., Iwasaki, T. and Arita, M. (1988) Seabed sand mining in Japan, *Marine Mining* **7**, 49–68.

Urabe, T. (1989) An overview of the seafloor hydrothermal mineralisation in the northwestern Pacific basin, Abs. Joint CCOP/SOPAC-IOC Fourth International Workshop on Geology, Geophysics and Mineral Resources of the South Pacific, Canberra, Australia, 24 September–1 October 1989, CCOP/SOPAC Misc. Rept. 79, Suva, Fiji, p. 110.

Uren, J.M. (1988) The marine sand and gravel industry in the United Kingdom, *Marine Mining* **7**, 69–88.

US Congress. Office of Technology Assessment (1987) *Marine Minerals*: *Exploring our New Ocean Frontier*, OTA-0-342 (Washington DC, US Government Printing Office).

Usui, A., Nishimura, A., Tanahashi, M. and Terashima, S. (1987) Local variability of manganese nodule facies on small abyssal hills of the central Pacific Basin, *Marine Geology*, **74**, 237–75.

Varnavas, S.P. (1986) An Fe-Ti-Cr placer deposit in a Cyprus beach associated with the Troodos Ophiolite complex: implications for offshore mineral exploration, *Marine Mining*, **4**, 405–34.

Volpe, A., Lougee, B. and Hawkins, J. (1986) Petrologic-tectonic evolution of the Lau Basin. EOS, *Trans. Am. Geophys. Union* **67**, p. 318.

von Rad, U. and Kudrass. H. (1984) Phosphorite deposits on the Chatham Rise, New Zealand, *Geol. Jahrb*, **D.65**.

von Stackelberg, U and Beiersdorf, H. (1987) Manganese nodules and sediments in the equatorial North Pacific Ocean, *Geol. Jahrb*. **D.87**.

von Stackelberg, U and the Shipboard Scientific Party (1985) Hydrothermal sulphide deposits in back-arc spreading centres in the southwest Pacific, *BGR Circular* No. 2, B.G.R. Hannover.

von Stackelberg, U and the Shipboard Scientific Party (1988) Active hydrothermalism in the Lau back-arc basin (SW Pacific) – first results of the SONNE 48 Cruise (1987), *Marine Mining* **7**, 431–442.

Webb, P. (1979) The development of the UK sand and gravel industry, in, *Offshore Mineral Resources*, Documents BRGM 7, Orleans, France.

Wiltshire, J. (1990) Platinum accumulation in cobalt-rich ferro-manganese crusts, *Proc. 4th Pac. Cong. Mar. Sci. Tech. (PACON 90), Tokyo*, Vol. 1, 405–12.

Windom, H. (1976) Lithogenous material in marine sediments, in *Chemical Oceanography 5* (eds. J.P. Riley and R. Chester), Academic Press, London, pp. 103–36.

Woolsey, J.R. (1976) Neogene stratigraphy of the Georgia coast and inner continental shelf, Ph.D. dissertation, Univ. Georgia.

Yim, W.W.S. (1979) Geochemical exploration for tin placers in St Ives Bay, Cornwall, *Marine Mining*, **2**, 59–78.

Index

Page numbers for figures are bold, those for tables are italic.